Florida A&M University, Tallahassee
Florida Atlantic University, Boca Raton
Florida Gulf Coast University, Ft. Myers
Florida International University, Miami
Florida State University, Tallahassee
University of Central Florida, Orlando
University of Florida, Gainesville
University of North Florida, Jacksonville
University of South Florida, Tampa
University of West Florida, Pensacola

– Science and Literature –

Bridging the Two Cultures

David L. Wilson and Zack Bowen

University Press of Florida

Gainesville · Tallahassee · Tampa · Boca Raton

Pensacola · Orlando · Miami · Jacksonville · Ft. Myers

Library of Congress Cataloging-in-Publication Data
Wilson, David L., 1943–
Science and literature: bridging the two cultures / David L. Wilson
and Zack Bowen.
p. cm.
Includes bibliographical references and index.
ISBN 0-8130-2283-5 (alk. paper)
1. Literature and science. I. Bowen, Zack R. II. Title.
PN55.W55 2001

801—dc21   2001048075

The University Press of Florida is the scholarly publishing agency
for the State University System of Florida, comprising Florida A&M
University, Florida Atlantic University, Florida Gulf Coast University,
Florida International University, Florida State University, University
of Central Florida, University of Florida, University of North Florida,
University of South Florida, and University of West Florida.

University Press of Florida
15 Northwest 15th Street
Gainesville, FL 32611–2079
http://www.upf.com

To Lindsey Tucker, Peggy Wilson, and Ron Wilson

Contents

Preface

This book is intended for those who are fascinated by the entire range of human knowledge and particularly interested in issues at the interface between science and humanities. Such a conjunction has always been a part of scientific and philosophical literature, and today the philosophy of science remains an important component of the academic curriculum, even as our specializations take us into ever more idiosyncratic research. We find that a number of texts called "Science and Literature," with various subgenres attached by colons to the titles, have become almost as opaque to nonspecialists as the literature of the specialties represented. Our book represents an attempt first to understand the basic tenets of science, and then apply them to more recent and particularized fields and results of scientific speculation and experimentation. While the logic of understanding its counterpart, literature, is far less programmatic than it is intuitive, we hope to show how a humanist might demonstrate another way of reasoning in a discussion of specific issues.[1]

What is radically new, experimental, and virtually without precedent is that our book is offered in the form of a dialogue. It represents two distinct points of view—two cultures—trying to see similarities in the questions we pursue and what we have in common as well as what our differences in methodology and thinking are. We are talking about the cultures we represent in order to understand better the kinds of answers we make to fundamental questions. When we speak of science and literature, we are not merely theorizing abstractly, but talking about ourselves and how we think, as individuals as well as practitioners of our diverse disciplines, and how we can aid each other in the pursuit of knowledge. The dialogue is everything.

We believe that the topic of interdisciplinary studies is exceptionally timely and is perhaps the most important subject in current discussion of the academic curriculum all over the country. While such discussion stems from an ageless underlying curiosity about the nature as well as specifics of human knowledge in general, the current debate and concern are associated with the increasing fragmentation of disciplinary knowledges and methodologies and the uncomfortable feeling that we are no longer talking to our colleagues across the disciplines. While most thinking people would hardly care to admit total unconcern about what colleagues in other fields are about, or on what basis they are proceeding with their work, the task of negotiating some basis of common understanding seems beyond many of us.

We seek to delineate the methodologies or concerns of our separate disciplines, and then to explore the use of scientific concepts and ideas in particular literary works, and, going on from such explorations, the two of us exchange views on some of the more controversial topics that bridge the two cultures, including possible limitations on scientific knowledge, the question of free will, and postmodern views of power, knowledge, and language. Specific texts, ranging from Genesis to *Brave New World,* are used to raise and illuminate more general issues concerning human nature and the ability of science to understand it.[2] We attempt to provide enough background in science for a meaningful analysis of the works from literature as well as for an exploration of the interface issues. Our analysis leads to an expanded discussion of some of the more compelling issues of our time. We come to the task, each with a deep love of our own discipline, and an interest in and respect for the importance of the other's. This respect seems to have been lacking in some participants in the recent "science wars." We share a belief that both science and humanities have important contributions to make to ideas concerning human nature.

The origin of our shared efforts was a course, "Science and Literature," that we recently taught together at the University of Miami. Writing a book about an experience we shared in a classroom involves choices, of voice, aim, audience, and scope. We wanted to retain in our prose some of the qualities of stylistic essence, opinions, and subject matter that characterized us as individuals, warts and all, and at the same time to employ in our prose the disciplinary vocabularies involved in our highly selective topics. The result was an approximation of the occasionally idiosyncratic tone of our lectures, discussions, and debates. To catch the spirit of the debate as well as the areas of agreement, the book presents our views expressed in our own individual voices (so labeled), and we have not attempted a merger, except for this preface. More details about the course that we taught together can be found in appendix A.

In writing a book for general readers interested in the issues arising from an analysis of the topics that bridge science and literature, we hope that our readers will derive some enjoyment from our attempt to show the ways in which these fields complement each other, and a source of insight into the thinking inherent in the scientific and humanistic cultures from our debates, when our views and fields appear to be in conflict. Through a selection of literary works, we highlight some of the practical uses of science in imaginative literature. And through an analysis of claims about science and humanities, we debate the limits of science in understanding human nature and arriving at truth.

Chapter 1 begins with a primary example of our different approaches to a common area of concern: how the sciences and literature deal with the origins of our environment and ourselves. Both science and literature have interesting things to say about where we have come from, who we are, and what we know. David explores the story of the origin of the universe and humans that has come from advances in science. That view is contrasted with Zack's discussion of two literary origin narratives—the biblical version in Genesis and John Barth's contemporary account of the "Night-Sea Journey" of an existential swimming sperm cell.

Chapter 2 provides a description of the methods of the natural sciences, a necessary foundation for our later discussions. This is followed in chapter 3 with a consideration of the state of the sciences dealing with humans, from psychology to history and evolution. That discussion, coupled with the humanist perspectives on knowledge and power, which take center stage in chapter 4, allows us to build toward an analysis of *The French Lieutenant's Woman* in chapter 5. Here we see the use of fiction to illuminate science, and of science to illuminate fiction. Our analysis of *The French Lieutenant's Woman* leads to issues such as uncertainty, determinism, and free will, which we tackle in chapter 6.

Chapter 7 builds the necessary scientific background in thermodynamics for understanding Pynchon's *The Crying of Lot 49*, which follows in chapter 8. That confronts us with such issues as the nature of reality, and our possible knowledge of it, which we explore in chapter 9, with Plato's "Allegory of the Cave" serving as the introduction for a scientific analysis of human perception.

In chapter 10, we consider scientific revolutions, explore their impact on what it means to be human, and ask whether there can be progress in science, given such revolutions. Chapter 11 serves as a literary/historical antidote to the idea of linear progress, as it explores the cycles of history, and their representation in several of Yeats's poems.

Chapter 12 confronts ethical and moral issues in science, an undercurrent earlier in the book that now is made explicit, as we prepare for the examination of *Brave New World* in chapter 13. Following a concluding summary chapter, we have appended the syllabus for our original course (appendix A) and a summary of the final debate among both teachers and class of a strongly stated version of scientific certainty, Rothman's *The Science Gap* (appendix B).

In retrospect, our book might seem to be an exercise in futility, owing to the enormous scope of the potential subject matter and the differences in approach of the disciplines we represent. By using examples from both science and literature to address various facets of common problems, we have projected the microcosm of our own remarks onto the macrocosm of attitudes and methodology each discipline might employ in understanding each issue. Like most of the people who work in interdisciplinary studies, we were looking for areas of commonality in the questions we asked, even as we explored the differences in the methods we used to frame our answers.

In showing how the sciences and the humanities approached universal problems, how as well as what the two areas thought about truth (and each other), we also made an effort, insofar as we were constitutionally able, to avoid unexamined generalities and platitudes, to demonstrate with specific examples, and to discuss a historical range of problems from the re-creation of ourselves and our environment, through human perception, to quantum mechanics and indeterminacy. In doing so we tried to account for the social sciences that breached the methodological gap between the hard sciences and the humanities, and the evolution of current disciplinary study. We also tried to elaborate the larger historical considerations with more recent examples of experimentation (literary as well as scientific) and theoretical thinking in a range of disciplines. No effort is made to gloss over the differences in the disciplines: the economic disparities between available funding, public perception of usefulness, and the certainty of arriving at underlying truth.

We offer this book not as any definitive answer to bridging the gap between these two disciplinary areas, but as a friendly and open attempt to give some idea of the scope, thought processes, and methodologies used by a scientist on the one hand and a humanist on the other. The topics and problems considered are enormous, and can easily be complicated by readers' opinions regarding subject matter and approach as well as our individual generalizations on science and the humanities—we leave plenty of room for making unintended enemies. In this book we do not mean to summarize the whole of disciplinary thinking in either of the areas we represent, nor do we attempt to offer new disciplinary discoveries. It was meant to provoke thought, discussion, enjoyment, discomfort, debate, and finally a measure of understanding. Ultimately, we hope to help begin the process of bringing two seemingly different enterprises together in a common search for knowledge and understanding of the human condition. The subject is too important to allow petty squabbles to interfere with the advances that cooperation might bring.

Acknowledgments

We thank the students in our class for their comments and questions, which served to clarify what we have written. For their most useful suggestions and corrections, we thank particularly Michael Gillespie, Tom Rice, and Margaret Wilson.

Primary Literary Texts

John Barth. "Night-Sea Journey." In *Lost in the Funhouse*. New York: Anchor Books/Doubleday, 1988, 3–13.

John Fowles. *The French Lieutenant's Woman*. New York: New American Library, 1969.

Michel Foucault. "Two Lectures," *Power/Knowledge: Selected Interviews and Other Writings, 1972–1977*, ed. Colin Gordon. New York: Random House, 1980, 78–108.

Aldous Huxley. *Brave New World*. New York: Harper and Row, Perennial Library Edition, 1989.

Plato. "The Allegory of the Cave," chapter 25 of *The Republic of Plato*, trans. Francis MacDonald Cornford. New York: Oxford University Press, 1965, pp. 227–35.

Thomas Pynchon. *The Crying of Lot 49*. New York: Harper and Row, 1966.

Origins

Wilson

We will begin our examination of science and literature at the beginning of the universe. What do we know about where we came from? What has been the history of the universe in which we find ourselves? I'm going to present the current scientific view about this, and then Zack will give us some of the perspectives of literature.

During the twentieth century we have made dramatic progress toward answering these questions. Let me tell you what we think we know and then go on to explore some of the weaknesses in our knowledge. Some billions of years ago, most scientists believe, there was a "big bang," as all of the energy in our universe, compacted in a small volume, began expanding. Scientists disagree about the exact time this happened, but ten to fifteen billion years probably is a reasonable range for the age of the universe. The most recent evidence suggests twelve billion.

The universe has been expanding ever since. We can see the expansion by examining the light coming from other stars and galaxies. The galaxies are spreading apart with time, as would be expected from a long-ago explosion at a single point. Within the first billion years after the big bang, stars and galaxies began forming. Galaxies are clusters of billions of stars, and there are hundreds of billions of galaxies.

Our earth and sun are in one of these, called the Milky Way. We are out on one spiral arm, nowhere near the center, which probably is a good thing, since there may be a "black hole" there, a place of high energy density, no life, and no escape. Our Milky Way is a rather typical galaxy. Those of us unfortunate enough to live in big cities cannot just look up and see why our galaxy is called the Milky Way. But on a dark, clear night, away from city lights, if one looks up at the sky one can see a band of stars more closely packed together than they are in the rest of the sky. The stars in our galaxy are clustered in that band, which can have a milky look to it.

Many lines of evidence support the view I have just sketched of the history of our universe. I will describe two. We can determine, by measuring the "red shifts" of spectral lines, how fast other galaxies are moving away from us. As stars in the galaxies recede at higher velocities, the color of light from the stars shifts toward the red end of the spectrum. The different speeds of recession from different galaxies, measured by their red shifts, give us evidence of a continuing expansion. Some of the light from galaxies now reaching us left the galaxies billions of years ago. In a sense, this is a way that we can look back in time. The orbiting Hubble telescope has enabled us to see light that appears to have been emitted more than ten billion years ago from galaxies that were forming just after the big bang.

There also is a background microwave radiation, an echo of the big bang, that was found at the wavelength predicted by the big-bang model. The discovery of this radiation helped to convince most astronomers that the model of a big bang, followed by an expanding universe, was the right one.

Within three minutes of the big bang, such elementary particles as protons, neutrons, and electrons were first formed. The energy density did not allow them to form until the expansion had gone on for a couple of minutes, and things "cooled" a bit.[1] These elementary particles are the building blocks for atoms and hence for life. It is rather remarkable that we can say something about what happened so many billions of years ago. Our knowledge comes from high-energy physics—studies of elementary particles in accelerators. This is an example of the interdisciplinary nature of science—that knowledge of what happened billions of years ago can come from studies of elementary particles in accelerators. I have described the view of the origin of the universe currently accepted by most scientists. Just fifty years ago, there were two competing hypotheses of the history of the universe. The big bang was one and the other was something called the steady-state hypothesis. It assumed that if you look around the universe now, and you look again in a billion years, and again in another billion years, things would look pretty much the same—the density of matter and galaxies would remain the same. But then existing evidence suggested that galaxies were moving away from one another—that the universe was expanding. Those who held to the steady-state view tried to adjust to such data by suggesting that matter was being created all the time, throughout the universe. Such a creation would compensate for the expansion, but this view of the creation of matter would violate the first law of thermodynamics, concerning the conservation of energy. As evidence grew for the big bang, supporters of the steady state dropped in number, then dwindled to a few. Their case became weaker and weaker with new data and analyses. All data and observations have supported the big-bang view and not the steady-state view. That is how science often works—through tests of competing hypotheses, comparing the predictions of the hypotheses with what is observed.

Let's work our way from the big bang toward humans. As I indicated above, elementary particles came into existence very shortly after the big bang. Elementary particles get together to form atoms, which compose all living organisms. Where did the atoms come from? One of the abundant elements, or kinds of atoms, in living organisms is hydrogen. A hydrogen atom, in its simplest form, consists of one proton plus one electron. All other atoms have more protons in their nuclei. Almost all the other elements that are necessary for life—such as carbon, oxygen, and nitrogen—were generated by nuclear fusion reactions in stars. Stars are giant nuclear furnaces that glow for billions of years, and sometimes cough out their products as they die. Stars generate energy by fusing smaller elements, such as hydrogen, into larger ones, and as they do, energy is released in the form of heat and light. Stars are thought to die because they run low on their primary fuel, hydrogen.

The first stars formed after the big bang could not have supported solar systems containing life since there was not enough of such elements as carbon, oxygen, and nitrogen present in the universe at that time. But as the first generation of stars exhausted their hydrogen fuel and exploded into supernovas, they spread a whole host of larger atoms around the universe. Our sun, which is a second- or third-generation star, then could have planets around it full of such heavier elements, and thus be capable of supporting life.

At the time that our solar system formed, about 4.5 billion years ago, there still was plenty of hydrogen around to condense and become part of the sun's furnace. Earth and the other planets contain a variety of the elements which were formed in the earlier, first-generation, stars. Thus, as Carl Sagan claimed, we are star stuff—the matter we are made of was forged in intense plasma furnaces in long-extinct stars, whose temperatures and pressures allowed for the merging of small nuclei into larger ones.

The same fusion events that occur in stars also take place when a hydrogen bomb is exploded. In contrast, atomic bombs are somewhat different. They gain their energy from the splitting, or fission, of heavier elements rather than fusion of smaller ones. In both cases the released energy is coming from matter, following Einstein's famous formula, $E = mc^2$.

Within a billion years of the formation of our sun and planets, life existed on earth. We can see the fossil remains of early living matter. Life most likely began spontaneously. Today, we can recreate primordial conditions and generate some of the building blocks of life—amino acids and such—in short order. We can play early natural-selection games with macromolecules, such as RNA, which may exhibit some self-replication, or self-copying ability. But details of the origin of life are not worked out. Interestingly, we know less about the origin of life than we do about how the universe began. There are continuing debates and hypotheses about life's origin—protein first, RNA first, and so on.[2] The state of this science is more like the state of cosmology earlier in the century as debates between the big-bang and steady-state models were going on.

As life came into being, cells formed, and all living matter today is composed of cells. A billion years or two after life's origin, some of the cells got more complicated, with internal membrane systems. Then some cells began to draw closer to each other: apparently there were benefits to growing in clusters, with the activities of some members of the cluster becoming specialized. After another billion years, life had moved from sea to land, and there were plants and animals around. For over a hundred million years, dinosaurs dominated, while a few frightened mammals scurried around underfoot. Then, about sixty million years ago, something happened to disrupt the environment across the earth; probably a big meteor hit the Yucatan area and dusted off the dinosaurs. Some mammals survived, and they began to flourish.

Primates appeared about half-way between the disaster and the present, about thirty million years ago. Around six to eight million years ago, probably somewhere in Africa, an ancestor of humans diverged from other apes. (Chimps are not our ancestors; we both diverged from a common ancestor about six million years ago.) Between two and four million years ago (depending upon how picky you want to get), more humanlike species evolved, but their brain size was perhaps one-third of ours—more chimp-sized than human-sized. Our knowledge as we go forward from here is incomplete at present. There probably was more than one line of hominids that split off, a branching of the tree of life, leading eventually to several species of *Homo*, such as *Homo habilis* and *Homo erectus*. All of these reached dead ends, except the twig that became *Homo sapiens*. *Homo erectus* appears to have emerged from Africa to populate Asia and Europe and remained in existence from about 1.8 million years ago until it became extinct about the time *Homo sapiens* appeared. There may have been some overlap: we may or may not have evolved from *H. erectus*. There certainly was overlap with Neanderthals, who were larger than we, and appear to have been present on earth as recently as 30,000 years ago. We may or may not have interbred with Neanderthals; this is the sort of uncertainty that characterizes our knowledge at present. It is interesting to note that Neanderthals were larger than we in both body and brain, since we often think that we survived because we were large-brained hominids. The evolution of mind among us and our ancestor species is still subject to much speculation.

On the tree of life, the *H. sapiens* twig is only about 225,000 years old. There are two views today of how modern humans arose and spread around the world. One is that we emerged independently in several places around the globe from something like *H. erectus*. The other, which is looking more likely as evidence grows, is that we emerged out of Africa about a hundred thousand years ago and spread—perhaps wiping out *H. erectus* and Neanderthals as we went.

This out-of-Africa view suggests some interesting things about genetics and race that may prove useful when we later consider the controversies surrounding IQ and the inheritance of intelligence. Humans are not all that genetically diverse as a species, but it appears that there is much more diversity if one looks throughout Africa than there is in, say, the whites of Europe and the Americas. That would make sense if it were just a selective subgroup that left Africa a hundred thousand years ago. Such a group would have been likely to have only a subset of the gene-pool diversity found in all Africans at that time.

This quick sketch of human evolution is oversimplified in some ways. We'll return to all of this as we consider whether our cultural basis for human races has any reality in the actual genetics of humans, and as we consider the genetics of intelligence.

This may be where we have come from. Where are we going? I wish I could say something definitive about where human culture is heading, but that is more challenging than attempting to project the really long run. Our own sun has a few billion years left before it runs out of adequate supplies of hydrogen fuel, so any concern about our sun's dying would not seem very pressing to a species that has only existed for about 225,000 years, but I happen to think that the issue of what will happen to the universe in the future is one of the big questions still facing science.

Several billion years from now, as our own sun begins to run out of hydrogen fuel and die, it will start to expand and reach a size that will engulf the earth. Obviously, if there are thinking beings still around, they will have to do something if they want to survive. One, possibly fortunate, coincidence, is the expected collision of our Milky Way galaxy with the Andromeda galaxy at about the same time. Perhaps passing stars in the Andromeda galaxy will give any still-thinking earthlings other options for habitable planets to replace a soon-to-be-cooked earth.

Over the longer run, there are several possibilities for the future of our universe. If the universe has enough matter in it, gravitational attraction eventually will cause the expansion of the universe to slow, and reverse, resulting in a "big crunch" as all matter collapses back together. What would happen after that? Might we have an oscillating universe, with another big bang following the big crunch? If so, each new bang would bring a new possibility of life arising again somewhere in the new universe. No remnant of the present one would remain, however. Nothing human would survive the big crunch, but the possibility of thinking creatures arising again would be there.

In the alternative scenario, if the universe has insufficient matter for a collapse, it will keep expanding; things will get colder; and stars will begin to form less frequently. Eventually the lights will go out—there will be no stars; everything will get dark and cold; and life will no longer be possible anywhere in the universe. Recent evidence suggests this second scenario as more likely.[3] Evidence indicates that there is not enough matter in the universe for a collapse, and that the expansion of the universe may be accelerating rather than slowing down. Thus, the idea that the universe will just continue to expand seems to be supported. That's life, or rather the end of life, unless there happen to be other universes.

It would, at least to me, be more comforting to think that there would be the possibility of some sort of conscious life going on beyond a few billion years; there is something so marvelous about a part of the universe, our brains, knowing about itself and having experiences. The universe would be a much duller place without it.

In sum, we now believe that protons, neutrons, and electrons appeared just after big bang; that the atoms in us, other than hydrogen, came mostly from stars; and that we evolved as one twig on the primate limb of the mammal branch of the tree of life. We will explore the mechanism underlying that evolution in a later chapter.

Does any of this disturb you? If you have faith in a God, does this scientific view leave any room for Her/Him? If not, do you toss all of this science in the trash bin? Or, do you throw the idea of God in the trash bin? Perhaps you try a little mix-and-match? Do you accommodate to this big-bang stuff—just stretch the seven days of creation to a few billion years? But what about this evolutionary monkey business? That certainly has been harder to take for some, ever since Darwin.

One obvious limitation to this scientific history of the universe is that we have no knowledge of what happened before the big bang. Is this where God fits in? Will our picture ever be more complete than "just" twelve billion years back? Some physicists argue that the kind of concentration of energy that existed at the time of the big bang precludes any detail being known of what came before. Time will tell.[4]

The above, brief history of life and the universe may have caused you to fall off of your chair the first time you heard it. Even just 150 years ago, we had hardly any knowledge of how we came to be; just fifty years ago much was still uncertain. Now, many of the earlier stories of origins have been turned on their heads, have been made into myths in one sense of the term—a sense that can make Bowen's blood boil. For some, this "News from the Science Front" may have been very disturbing. We now actually know something about how we have come to be here in the universe, at least in terms of how humans got started—I don't mean to say that we yet understand the details of how human minds coevolved with human language and culture. We don't know everything; and while some of the details probably will change, we don't have to speculate in a vacuum about origins anymore. I would predict that, in broad outline, no radical change due to scientific advances is going to change this description in the foreseeable future!

"How can you have any sense of how likely these scientific views are to change?" you might correctly ask. There have been radical changes in scientific views in the past, and we will explore such revolutions in a later chapter. Perhaps in the fine tuning something will turn up, an anomaly, that will lead us to a different conclusion. While such a thing remains a remote possibility, some of our more recent revolutions and observations have only reinforced aspects of this account, and the knowledge base of the natural sciences became much more complete during the twentieth century. The interconnections and cross-confirmations among different sciences lend considerable support to the sense that the general view we now have of the history of the universe is not too far off.

That is science's story up to now. Before you can judge how much confidence you should put in the history of the universe that I have outlined, it probably will be worthwhile for you to know something about how this scientific knowledge has come about. What kind of method was used to make these discoveries? How reliable is the knowledge? Can we trust it to give us accurate, or at least moderately accurate, descriptions of the world? We will move on to explore such questions as we examine the scientific method in the next chapter.

As I mentioned above, this scientific story of the history of the universe has disturbed some who come with particular religious or cultural views. Others are uncomfortable that science, and more specifically the technology that has been derived from it, has come so to dominate our culture. Some people charge that science is leaving out what is important in making humans human, such as our minds, morals, and arts. Different areas of science exhibit different levels of certainty, often because of complexity or the newness of the area. I am thinking here especially of the human sciences, and we will explore these issues in chapter 3.

Bowen

In many ways science and literature are not so different. Both rely on metaphors, such as the big bang, denoting an explosive image that carries connotations of a Schwarzenegger film on the one hand and sex on the other. In order to facilitate comprehension, scientists often couch their descriptions of the physical processes in terms of narratives, and thus concede their reliance on the techniques of humanistic methodology as well as scientific thinking. But scientific explanations often suggest the writers' conviction that they are disclosing unbiased truth in describing the processes of the natural universe, so that in this view, narrative and truth are coincidental. Except for religious dogma, narratives are valued by contemporary humanists not solely for their claims to truth, but also for their creativity, for the aptness of their message in answering larger questions, and for the aesthetics of their formulation.

Like scientists, we in literature, and the humanities in general, describe phenomena such as origin theory in metaphoric terms. For me, at least, science's big-bang theory is attractive because I see it less as a representation of the "truth" than as an attractive metaphor, with all the apocalyptic quality of Armageddon and the possibility of a salvific resurrection and survival for life in general. The questions and concerns of the students in our class regarding both the explosive beginning and implosive end to life, along with the hope of a new explosive beginning carry a heavy religious freight that makes a eulogy of the narrative. Despite its remoteness, the explosion/implosion process and its apocryphal overtones provoked in the students an immediate emotional concern for life, a concern made no less real by such lesser but equally disturbing predictions as the global-warming scenarios of scientists now being realized around the world. Nobody I know associated with the humanities today is immune to placing a good deal of prudent credence in scientific belief. In literary study, however, narratives are an aesthetic and intellectual end in themselves, and most of us in the humanities have long since ceased to regard them as unassailable truth no matter what their origins. So, I choose to represent the problem of the origins of life, and particularly human life, in terms of two possible narratives.

It is safe to say that even the earliest people provided themselves with metaphoric explanations of how they got there, and these have undergone a series of increasingly diverse tribal mutations in a continual evolutionary flux that stretches to the present day, with ever fainter, if still salient, echoes of their linear descent. I would like to investigate two origin narratives, one classical and one contemporary, to afford some idea of the range of possibilities over the span of a millennium or two.

The first is the traditional Genesis narrative, one which answers a number of questions even as it creates others. Genesis is, of course, not the first, last, or only word on the subject. The origins of the Genesis text in earlier Middle Eastern incarnations reflect both the diversity of its claims and answers to problems raised by earlier polytheistic religions. There was political as well as spiritual motivation behind a narrative establishing a patriarchal monotheism that embraced the history of the Jewish people. Genesis underwent an evolution between the time of its oldest, Jehovistic source (J) in the tenth and ninth centuries B.C. through the Elohistic (E) version in the eighth century to the priestly writer's version (P) in the sixth to fifth centuries. These principal versions were in turn influenced by preexisting religious mythologies, such as the Mesopotamian associations of watery origins and seasonal patterns arising from that civilization's location between the Tigris and Euphrates rivers; Sumerian myths concerned with the primeval sea; Greek-derived deities like Uranus (sky god) and Gaea (earth goddess); and other cosmogonic myths culminating in the creation of mankind. Most usually depict the earliest era of the world as being closest to perfection, and an eventual progressive degeneration as it grows more distant from the original creative impulse. All of these find echoes in Genesis.

Among the several references to polytheism that survive in the Bible are those that exist between other angelic immortals and God's new human creations. And there are of course two different versions of the creation itself: in the first chapter of Genesis God simply creates human beings in his own image, bidding them to "Be fruitful and multiply, and replenish the earth, and subdue it" (1:28). The second narrative, in Genesis 2:7, has God fashioning Adam out of the dust of the ground, and Eve from Adam's rib. Adam is given dominance first in the power to name everything, including woman, and secondly as the source of Eve's existence, establishing a gender-based political hierarchy later to be ratified in Eve's heightened susceptibility to temptation. Scholars have seen the subjugation of women as linked to the establishment of the patriarchy coevally with creation itself, citing the older, often female-dominated, religious mythologies as something to be discouraged by creative fiat.

Starting with the male genderization of God himself and then loading Eve down with the primary burden for original sin combined sexism with a second concern of the Hebrew patriarchs, monotheism. This proscribed the worship of multiple deities (many of them female), and in the process nullified systems of worship, regulated often by priestesses, in competition with their Hebraic male counterparts. In Genesis the symbolic animal incarnation of evil is a snake, in its earlier incarnations represented in the circular form of a uroboros devouring its own tail, as a circular feminized symbol of life's processes.[5]

Even in their insistence on monotheism the patriarchs were unable to eliminate other male immortals such as Satan and the angelic host. Further, God's hint that acquiring the knowledge of good and evil leads to an unwholesome immortality—"Behold, the man is become as one of us, to know good and evil: and now, lest he put forth his hand, and take also of the tree of life and live for ever" (3:22)—expresses a tacit fear of an expanding polytheism. In the light of its evolution from many such diverse sources, Genesis was a difficult story to subject successfully to the biblical cleansing that took place throughout its scriptural history.

There are far fewer collective politics in the second, contemporary, creation narrative I will consider. "Night-Sea Journey" is the second story in John Barth's 1968 collection *Lost in the Funhouse*. The first "story" in the volume, "Frame-Tale," consists of a strip of paper with the following sentence fragment printed seriatim on the front and back of the first page: "Once upon a time there / was a story that began." The strip was meant to be twisted once and joined at the ends in one figure-eight strip (a Moebius strip) so that the text forms a continuum infinitely repeating itself and providing a message to be read from an ever-changing point of view. The "Frame-Tale" mirrors the recirculating structure of the book itself in which the last story is a prelude to the first and so on. In the last story, the protagonist, living in the age of Homer, and marooned on an Aegean island, writes his life's adventures on the skins of a diminishing and finally extinct herd of goats, and launches his self-referential manuscript out to sea in a succession of the large wine containers called amphorae. He deposits some of his sperm in the jugs to complement or metaphorically equate with his written record. The relation of the amphorae/sperm to the persona of the first story confirms the interpretation that the troubled existential persona of "Night-Sea Journey" is really a sperm cell swimming up the uterine track to complete the reproductive sequence and begin life all over again, as with the ever-repeating Moebius strip.   This origin story borrows a lot from science as well as from philosophy and from Genesis. Its environment is watery; the sperm cell has a tail and swims; and evolution and gynecology inform the text as much as do existentialist philosophical questions regarding the utility of life. The sperm has all the reasoning capabilities of a modern mature man, wondering about the tragedies inherent in the deaths of untold millions of his fellow beings, brooding on the waste of effort in survival, on the futility of life's existence, on the motivation for undertaking the arduous journey to the egg, and on the meaninglessness of the whole enterprise.

In Barth's story the role of the ovum, the ultimate attraction of which drives the sperm/swimmer's effort—in short the role of womankind—is of ultimate importance. Unlike Genesis, this narrative does not attempt to rationalize or assert any male gender supremacy, or to curtail or elimi-nate the female aspects of procreation, but neither does it attempt to make women articulate or understandable in human terms to the strug-gling sperm. The egg is never heard from and remains the mysterious attraction, spiritual in its subtlety, silent on its human concerns—the means and the goal of the entire process of life. Conversely, the role of the Maker—in this story the father/prostatic provider—is, through procre-ative gender identification, undeniably a male god, and the capitalized "Heritage"—which I take to mean human experience as well as DNA—the problematic result of the journey. We are offered a view of male speculation, which begins as does Genesis, before the full impact of the female principal, already apparently existing in the minds both of the lusting Maker and of the "Maker's Maker," is realized.

In his adaption of Judeo-Christian ideology, Barth leans heavily on the motivational quality of love:

> "Oh, to be sure, 'Love!' one heard on every side: 'Love it is that drives and sustains us!' I translate: we don't know *what* drives and sustains us, only that we are most miserably driven and, imper-fectly, sustained. *Love* is how we call our ignorance of what whips us, 'To reach the Shore,' then: but what if the Shore exists in the fancies of us swimmers merely, who dream it to account for the dreadful fact that we swim, have always and only swum, and con-tinue swimming without respite (myself excepted) until we die? Supposing even that there *were* a Shore—that, as a cynical compan-ion of mine once imagined, we rise from the drowned to discover all those vulgar superstitions and exalted metaphors to be literal truth: the giant Maker of us all, the Shores of Light beyond our night-sea journey!—whatever would a swimmer do there? The fact is, when we imagine the Shore, what comes to mind is just the opposite of our condition: no more light, no more sea, no more journeying. In short, the blissful estate of the drowned."(5)

Thus Barth offers a narrative of origin that is also a tale of conclusion, a continuum of the life/death process refined by thousands of years of human speculation, the whole related in the guise of postmodern fantasy, heavily influenced by Schopenhauer's description of "will" as "a blind creative urge essentially destructive and chaotic except where intellect can make some partial sense of it."[6] While the sperm's questions are ancient, the journey account depends on a scientific microscopic description of a sperm's tail/tale, a trip made by literally millions of cells paddling most of the way against current and odds, with an ultimate normal survival rate of less than one in a million. The metaphors of modern science are carried to their ultimate conclusion in a statement of humanistic perplexities, the watery birth itself coinciding with Darwinian theory even as the little, if complex, fish is about to begin its evolution through all the embryonic stages of the history of human evolution into a fetus and beyond. The issue of where we come from is only a segment of the circular concept of life represented in Greek and other religions, traveling again and again across waters to and from some sort of Elysian fields on the other shore, before being resurrected and submerged in the river Lethe's forgetfulness.

"Love," or its linguistic surrogate in the Bible, is the guiding, resurrecting, and procreating word in Barth's story also. Although the swimmer in Barth's story frets himself into solipsistic despondency, he cannot help completing his journey, as he hears in his mind her siren song: "'Come!' she whispers, and I have no will . . . 'Love! Love! Love!'" (12–13). Although a number of feminists considered the early Barth works chauvinistic, he certainly restored the preeminent importance of women in creation narrative.

The overall point in introducing such diverse stretches of the human imagination is to describe by example the narrative process as a way of examining life, specifically human life, for which all forms of knowledge ultimately seek explanation. By providing an avenue for understanding the nature and methods of coming to grips with common problems, whether they be political, social, or experiential, human intuition plays as great a role as scientific verification, especially when it comes to such cosmic issues as origin and even conclusion theory.

Wilson

Zack's comments about the big-bang model deserve a little postscript. This term for the expanding-universe model was actually first used by a supporter of the competing steady-state hypothesis, which I discussed earlier in the chapter. His attempt at derision backfired when the term became popular.

Zack writes of those in literary studies not accepting "unassailable truth" no matter what its origin. I can only add that most scientists would agree with this view, as I hope to convey in our next chapter on scientific methodology.

How to Do Science

Wilson

I'd like to give you an overview of what the scientific method is and how it works. It's easy to get lost in the details, so it will pay us to keep things in perspective—to see the forest as well as the trees. How do we think scientifically? What sets science apart from other intellectual endeavors? Are there special procedures that make a field scientific? Why isn't astrology or "creation science" considered scientific? In what follows, I'll focus on the natural sciences and save the social sciences till later.

My wife's Uncle Milt said to me one morning some years ago, "Dave, that vitamin C really works against the common cold." Now, to Uncle Milt, Linus Pauling, the double Nobel laureate, was a hero, so, when Pauling suggested that large doses of vitamin C would help to prevent the common cold, Uncle Milt started taking it. Milt went on to say, "I've been taking vitamin C every day now, and I've had only one cold this winter." Given that we both spent that winter in Chicago, he had a right to consider it an accomplishment. "But Milt," I said to him, "I haven't taken any extra vitamin C this winter, and I haven't had *any* colds."

This exchange highlights two problems with Uncle Milt's assertion about vitamin C therapy. First, he did not have a "control group." Second, his number of subjects, namely one, was much too small. That's why my single counter-example was enough to do in his observation. Neither his experience nor mine was adequate to determine whether or not vitamin C helps prevent the common cold. My observation simply pointed up the inadequacy of his as a basis for reaching the conclusion he reached.

I don't want to knock Uncle Milt's attempt to test a hypothesis. When we think hard about things, we often are thinking in a crudely scientific way. Science just gets more sophisticated, systematic, and careful about tests and the conclusions we can draw from them. Much of what we do as scientists has its basis in ordinary thinking about problems and attempts to find solutions for them. In science one of the things we learn—sometimes through the mistakes we see others make, sometimes through our own—is how to test hypotheses. Some tests are better than others.

Can we identify a process that defines what science is all about? There are some general descriptions of science that can help us grasp what it usually entails, but there is always something that will elude that grasp because science and the world tend to get pretty complicated. Let's take a simple approach first. As we do, you will see that there is as much art and creativity to science as there are set formulas and fixed ways of doing things.

I. A General Model

What sets science apart from so many other endeavors or forms of thinking?[1] The answer to this question concerns mainly the general method followed by scientists, a method that is drummed into most school children. You probably will remember from your own school years being taught something like the following:

1. Make observations.
2. Formulate a hypothesis (or hypotheses) and make prediction(s).
3. Test the prediction(s).
4. Confirm, reject, or reformulate the hypothesis.

This is sometimes referred to as the scientific method. The sequence, as presented to generation after generation of students, usually starts with observations. One tries to make sense of the observations. Making sense of observations can involve classifying them or trying to understand relationships among them. Even such preliminary activities involve the formation of hypotheses. A crucial action that tends to set the scientific enterprise off from many other kinds of intellectual activities is the subsequent step of testing the hypotheses and of using the results of such testing to determine the accuracy of the hypotheses.

The above list and description covers the essentials, but the list is not quite complete. Consider point 1. The making of an observation requires an observer, and any observer requires a system, a structure, for making observations. All such systems or structures involve assumptions. So, we do not simply start with observations. At the other end of the list, one does not make just one test and then accept the hypothesis. In reality, science engages in an ongoing process of testing and refinement. Therefore, a more complete listing of the steps involved in scientific analysis would be more like the following:

1. Make use of axioms and assumptions.
2. Make observations.
3. Formulate hypotheses and make predictions.
4. Test.
5. Confirm (or reject, or reformulate) your hypotheses and predictions.
6. Make further tests of confirmed or reformulated hypotheses.
7. Repeat indefinitely.

There are other ways of describing how science is, or can be, done, but the listing above is a good starting point for us. Let's now consider each of the steps in some detail and go on to some further descriptions and elaborations.

Axioms and Assumptions

The axioms and assumptions on which scientific knowledge at first relied were based on our ancestors' brain structures, which came about during embryonic and infant development. These brain structures underlie our sensations and perceptions. What we see, hear, touch, taste, and smell are the result of the processing of information in our brains. The assumptions, including such basic concepts as our perception of space and our sense of time, develop as our brains develop, as genes interact with their environments during development. We can assume that each of us started, as children, with axioms and assumptions very like those of our ancestors who got this whole scientific thing going. But we have the advantage, through what we learn from history, of understanding the modifications in the axioms and assumptions that have occurred during hundreds of years of scientific advancement.

Thus, the initial axioms or assumptions of our ancestors were not fixed, but most or all of them were (and still are) themselves subject to testing and modification. This may seem contradictory until one realizes that an assumption for one experiment or test can be the hypothesis for another experiment or test. Furthermore, an experimental test can fail not because a hypothesis is wrong but because of an error in an assumption or axiom. Occasionally a scientist with an apparently failed hypothesis is wise enough to recognize a possible problem with an assumption, and proceeds to test its validity rather than discard the hypothesis. Of course, if the assumption gets modified, then it is modified for all of science. Some major assumptions have undergone revision recently. Einstein's theory of relativity gave us a new view of space and time, and quantum mechanics has radically changed our assumptions regarding both the particulate nature of matter and the issue of continuous versus discrete energy levels.

Let's take this last issue as an example of how assumptions about the universe can change. If someone were to tell you that it was possible for you to walk at a rate of one mile an hour or three miles an hour, but not at the rate of two miles an hour, you probably would think this was crazy. In everyday life, we assume that energy levels are continuous. If, however, one looks at energy levels in individual atoms, the story is quite different. According to quantum mechanics, an electron orbiting an atomic nucleus can only have certain discrete energy levels. Levels between these are never observed. We will explore this limitation in more detail later on, when we consider the structure of the atom. That this limitation existed was quite a surprise to physicists, and required changing a basic axiom or assumption that they had been making.

While scientific assumptions and axioms are not fixed, at any particular time there is an accepted set that underpins all that is done in science. Of course, for a particular area of science or particular hypothesis some of those assumptions are needed and others are irrelevant. There is no master list of such assumptions, and scientists in writing their journal articles do not normally explicitly list the fundamental assumptions they are making, or if they do list some, it will be a very partial list of immediately relevant, or possibly suspect, assumptions that they are making or testing. Scientists will, however, try to be careful to list any assumptions that are not among the group commonly accepted by others in their area of science.

Many assumptions, including some of the major ones, are so basic that they are only to be found at the beginning of introductory textbooks. For instance, in the field of biology today, the working assumption is that the laws of physics and chemistry underlie and govern all of the processes and events in living things. That point usually is made in the first few pages of most introductory biology texts. That assumption was not always made. Even as late as the end of the last century, a figure as preeminent as Louis Pasteur was a firm believer in "vitalism," the view that living things contained a "spirit" or "force" that functioned outside the laws of the sciences concerned with inanimate objects, such as physics and chemistry. Vitalism was once a very easy view to hold: we see big differences between living and nonliving things. We have an innate feeling that there is something very special about living organisms that is not shared by rocks or sand. It was not hard to imagine that the differences involved forces that went beyond those that govern inanimate objects.

Pasteur's belief in vitalism was one of the driving forces behind his attempt to demonstrate that there was no such thing as spontaneous generation of life. During the earlier part of the nineteenth century vitalism was so ingrained that many scientists believed that only living things could make the chemicals found in living things, the so-called organic compounds. It was thought that these chemicals in living things were not subject to the same laws as inorganic compounds. Then when chemists began to make such organic compounds out of inorganic compounds in test tubes, the mystery disappeared. Organic chemists use the same underlying physical laws and forces as inorganic chemists.

Similarly, biology in this century has made such great strides that one can describe many of the formerly mysterious properties of living things—growth, movement, reproduction, heredity, metabolism—on a cellular and molecular level.[2] Today, our hypotheses and experiments in biology rest on the assumption that biological science is built on physics and chemistry, and that the laws of physics and chemistry are adequate to describe living things and their processes. At the same time, that assumption is being tested daily. Anyone who could demonstrate an exception would quickly become a center of attention.

The ability to test assumptions and axioms is a very fundamental characteristic that serves to set science off from many other intellectual endeavors. The basis of scientific knowledge is not fixed. Many other belief systems have fixed axioms that cannot be questioned within the system. Scientific knowledge is built on a modifiable foundation.

Note that Pasteur was able to make significant advances in our understanding of disease even with what today is known to be an incorrect assumption. Pasteur's belief in vitalism gave him a bias against spontaneous generation of life in decaying matter, but the accuracy of his experiments did not depend upon the accuracy of his assumption of vitalism, fortunately for him.

This entire scientific process raises some interesting philosophical questions. For instance, in the pursuit of accurate, correct hypotheses, one might ask whether our forebears' axioms and assumptions limit us in some unavoidable way. In short, can we get from here to there? Is truth accessible from our starting point? Despite such open questions, we are playing the game of science today, and it appears to be working, and may be the best game in town.

Observation

On the basis of the current axioms and assumptions, we make observations. For instance, we observe that the sun appears in the east every morning—at least when it isn't so cloudy that sure observation is impossible except by airplane, satellite, astronaut, or by moving to sunny Florida! Making observations seems easy, but it can be tricky. We must be careful to observe well. We also must be alert to events that we may not have thought relevant, but could prove to be more interesting and important than what we are studying—we must be prepared for serendipity. That is how penicillin, the first antibiotic, was discovered. The microbiologist Fleming was examining agar plates covered with bacteria. He found a mold growing on one of the plates, a contamination that normally results in a quick tossing of the plate into the trash. Fleming happened to notice a halo of lack of growth around the patch of mold. He concluded that something produced by the mold might be inhibiting the growth of the bacteria. Thank goodness he noticed.

Max Delbruck, a scientist I knew at Caltech, used to suggest the need for "limited sloppiness" in scientific research. He suggested that a little variation in how one does an experiment could actually be a good thing because, if it made an unexpected difference in one's results, it could lead to a new discovery, such as Fleming's.

Even so-called "simple observations" are not always so simple. Observations can be in error, and always involve interpretation and assumptions on the part of the observer. To give but one personal example, I often use pipettes to deliver specific volumes of solutions to test tubes. Pipettes are marked by their manufacturers, and a particular one that I was using was marked to hold 5 milliliters (ml), with gradations each tenth of a milliliter. I happened to be dispensing the liquid from the pipette into a similarly graduated cylinder. I had loaded the pipette to 4 ml, and just happened to notice that the volume in the graduated cylinder indicated only about 3 ml! Further testing showed that the pipette was in error. At its 5 ml mark, the pipette contained an actual volume of less than 4 ml. One does not normally inspect or test pipettes. This one had apparently been made on a machine that had jammed at its smallest setting for inscribing volumes onto pipettes. A 20 percent error is quite considerable, and could influence a lot of experiments, and certainly would contribute to scatter in one's results. Even simple observations rest on assumptions.

Hypothesis and Prediction

On the basis of observations we form a hypothesis. It doesn't form itself, and producing one can involve more than a little creativity and insight. Some hypotheses are simple. One might be that the sun revolves around the earth once every year. People once thought that the earth was at the center of the universe. In fact, it was so firmly (and religiously) believed that, when Galileo supported Copernicus's view, that the earth was not standing still at the center of things, he was forced by the religious leaders of his city to retract his support publicly, on his knees.

Of course nothing limits us to just one hypothesis, and some of the best science is done when two or more hypotheses are formulated as alternatives to one another, and then a test is devised to distinguish among them. Another possible hypothesis relating to sun and earth is that the earth is rotating on its axis once every 24 hours and revolving around the sun once a year. If that is indeed the case, the other planets also should appear to revolve around the sun, rather than the earth. And indeed, observations indicate that the other planets do progress in elliptical orbits around the sun. That result was a challenge to the hypothesis that the earth is at the center of the universe.

More sophisticated or derivative hypotheses require that one first make predictions from the hypothesis. Then it is the predictions that are tested. For instance, the hypothesis might be that vitamin C will prevent the common cold. The prediction then would be that someone taking 100 mg of vitamin C each day will not get a cold. A test of that prediction might involve giving volunteers vitamin C and then seeing if they catch a cold. If they do, then the hypothesis or predictions will have to be modified. One might try a variant: "Vitamin C will reduce the *likeli-hood* of one's catching a cold." Or again: "It takes 500 mg per day, so try again at that level." There are almost endless possibilities, even with this seemingly simple hypothesis.

Totally refuting a hypothesis can take considerable effort in the design and execution of experiments. The case needs to be airtight enough to convince other scientists that the hypothesis is wrong. This includes satisfying most possible objections to the experiments or interpretation of the results. Also, many hypotheses can be reworded or reformulated in an attempt to get around the experimental results. However, if one does succeed in showing that a hypothesis and simpler reformulations of it are in error, the scientist usually will move on to other possibilities. If you don't on your own, the agency that funds your research usually will see that you do before long anyway!

Tests

Creating tests of hypotheses, or of predictions from hypotheses, can be challenging. One needs to design experiments carefully to keep unwanted variables from interfering in a "clean" test of the hypotheses, whenever possible. For example, you might be studying the life span of an insect and notice that there is a great variation in how long they live. Some groups die after an average life of a few weeks while other groups live for several months. Then you realize that it is in the winter that the insects are living longer, and you realize also that you had better hold the temperature more constant if you want to get reproducible results.

Ideally, one holds everything else constant while testing one variable. In practice, this can be difficult. It is easier in physics, harder in biology, and almost impossible in sociology. As it gets harder, one must devise other means of separating out the effects of different variables.

Any real test of a hypothesis or prediction must be powerful enough to refute the hypothesis—it must demonstrate that particular observations or analyses resulting from the test show the hypothesis to be wrong. Of course, if such observations actually are made, the hypothesis may indeed be wrong. On the other hand, it is often all too easy to design experiments that seem to confirm a hypothesis but actually are not a real test of it because there is no conceivable way for the test to disprove the hypothesis. One should design a test that has at least the possibility of falsifying the hypothesis.[3] One should use the test with the greatest potential of showing that the hypothesis is incorrect. The possibility of falsification has become a hallmark of the scientific method, and good science involves testing a hypothesis at its limits.

If one draws a couple of plastic turtles out of a bag, and they both are green, one might hypothesize that all plastic turtles in the world are green, and one might continue to test the hypothesis by pulling plastic turtles out of the same bag. If the next few turtles also are green, should one continue to test the hypothesis by taking plastic turtles out of that same bag? Or would it be better to try another bag? Or check out a plastic-toy factory? Or perhaps find someone who collects plastic turtles to see if they come in any color but green. Scientists sometimes can trap themselves by designing experiments that are the equivalent of looking only in one bag.

There is another side to the falsification view. Sir Karl Popper argued that some hypotheses were not scientific at all because there was no way that they could be falsified: that is, there was no conceivable observation that would show the hypothesis to be false. Such a situation can arise and not even be noticed at first. Popper took as an example the dream "theories" of Sigmund Freud. Freud had explanations for a great variety of dreams, as to their significance and hidden meanings. Popper claimed that Freud's "theories" are simply not scientific because there is no conceivable way of falsifying them. To be scientific, the theory should, for example, identify some dreams that humans cannot have. Should someone ever have such a dream, the hypothesis would be refuted. The chance of refutation should be there. Without it we don't have science, according to Popper.

As a practical exercise to drive home this point, consider a set of cards, each of which has a letter on one side and a number on the other.[4] A hypothesis proposes that whenever the letter A appears on one side of a card, the number on the other side of the card will be 3. Now consider four cards on a table, representing four possible "tests" of the hypothesis. The four cards have, respectively, A, B, 3, and 4 on them. Which two of these four actually serve as a test of the hypothesis "If A, then 3?" (Make your choice, then read on; no peeking).

Obviously the first one does. Should the card with an A on one side be turned over, and there is a 4 on the other side, then the hypothesis can be discarded as false. So, this is a good test of the hypothesis, and if there is a 3 is on the other side, it serves as a confirming instance of the hypothesis.

The second card does not constitute a test of the hypothesis because a card with B on one side can have anything, even a 3, on the other side without affecting the hypothesis. The hypothesis says nothing about what is on the other side of a card with a B on it.

How about the card with a 3 on it? This is the tricky one. A weak scientist could spend all day turning over cards of this sort. They would appear to confirm the hypothesis every time an A was found on the other side of such a card, and there might be journals and reviewers accepting the research for publication, if they didn't recognize the weakness of the test. But notice that there is no way that turning this card over can show the hypothesis to be wrong, no matter what letter is on the other side. It is not a test of the hypothesis.

The fourth card is a good test: if there is an A on the other side of the card with the 4 on it, then the hypothesis is wrong.

Even in very basic fields related to physics, there can be significant challenges to developing a good test of a hypothesis. For instance, it is not possible in astronomy to move stars around in the heavens, so one must look for "natural experiments" which may be occurring in the sky, if one knows where and how to look. One might need to look beyond visible light, at x-ray emissions, or at gamma rays. The test can sometimes become a search for a phenomenon, or its absence, as in the case of the background microwave radiation that served as a confirmation of the "big bang" hypothesis.

Experimental design is challenging in other ways, too. In some fields, like chemistry, microbiology, biochemistry, and molecular biology, one often uses mixtures of chemicals. I had to discard all of the results from my first six months of research as a graduate student because of an error in experimental design. I was studying bacterial viruses. The bacteria were grown in a medium containing a variety of nutrients, and after the bacteria had gotten dense, I would add viruses. These would infect the bacteria, multiply, and kill the bacteria in about a half an hour. But the results would vary when I attempted to repeat my experiments. Sometimes the infection would go well, and I would get a hundred or more viruses produced for each bacterium. At other times, the bacteria would die, but few viruses would be produced. As a new graduate student, I found this very frustrating, and it was not easy to track down. It turned out that halfway through the infection, the medium was running out of amino acids, which are essential for making the proteins that are a part of each virus. I was getting results that varied according to when during the infection the amino acids became depleted. Wasting six months of work made this a hard lesson to learn, but at least it was early in my career, so I've had years to benefit from the experience—I've learned to be more careful. I also learned that even very simple experiments can take a long time to get right.

Commonly, especially as science has grown more sophisticated, the test of a hypothesis requires an analysis of data gathered during an experiment. The analysis of data is necessary for most experiments, but carries with it the risk of introducing error and usually requires further assumptions.

For instance, in 1991 there was a report that the first planet had been discovered outside our solar system. A couple of months later the scientists involved retracted the result. They had made an error in their data analysis: they had failed to include the movement of the earth around the sun as a correction factor while they were looking at this more distant object. Quite embarrassing, but also quite honest of them to make the retraction. That is what science is supposed to be all about. Absolute honesty. If you make a mistake you let others know—better you tell them than someone else.

A more common example of the chance for error, at least in the biological and social sciences, comes from the statistical analyses that often are required. In such experiments a control group and an experimental group are compared. You might give vitamin C to one group and a placebo to the other. Then you count the number and severity of the colds that individuals get. You compare the results by a statistical test. The statistical analyses usually involve assumptions about the kind of distribution, or scatter, that characterizes the data. After making those assumptions, at the end of the analysis, one might be in a position to say that the odds of the result occurring by chance alone are one in a hundred or one in a thousand. Thus one concludes that the results are "significant." But note that one in a hundred or one in a thousand of such reports should be by chance alone. Many more than one hundred or one thousand such results are reported in the scientific literature every year. Some are bound to lead to incorrect hypotheses being accepted, at least temporarily, because of chance happenings.

Does this mean that we shouldn't report such results? No. Most of the observations are correct, that is, they are significant. Those that occurred by chance alone, if they are important, will be repeated and are not likely to be confirmed by the same scientist or by others in the field. That is one reason why we do need to repeat experiments and have others confirm our results. That need to repeat and reproduce results is not always recognized as important by some animal-rights activists, some congressmen who vote on bills for funding research, and others. Science will not work well without such repetition, and it also diminishes the likelihood of fraud.

Reformulation and Retesting

One can never prove things to be absolutely true with the scientific method. One often feels confident that something is likely to be true, or at least a very close approximation to the truth, but the uncertainties about assumptions and axioms, coupled with the possibility of mistakes in experimental design and interpretation, always keep one from being 100 percent certain. Suppose, for instance, that a large, well-controlled study were to show that vitamin C at 500 mg/day reduced the likelihood of getting a cold by more than 50 percent. The result is published, and everyone starts taking vitamin C.

Then someone notices that a particular brand of vitamin C does not prevent the cold. Further work shows that there was a contaminant in the other brands of vitamin C. Let's call it Substance Wilson, after the great scientist who first isolated it! More tests show that Substance Wilson is what reduces the likelihood of one's catching a cold. The hypothesis that we thought was correct was all wrong. Such things can happen.

Retesting must always be done, and new tests must be devised. That, again, takes creative thinking. Reformulation or refinement of hypotheses in the light of new data is not unusual.

From Hypothesis to Theory

The more confident that scientists doing related studies are, the more likely they are to call a hypothesis a theory. A theory is a hypothesis, or even a group of related hypotheses, that has stood unrejected after testing under a variety of conditions and circumstances by a variety of scientists. It has been tested, retested, and tested again, without need for rejection or reformulation.

If the theory has ever seemed to fail a test, then the test (that is, the design or execution of the experiment) has been found wanting. Otherwise the theory would have been rejected. The only (temporary) exception would be in transition periods, where a test, confirmed by others, indicates that a theory in its present form is not likely to be true, but another has not been devised to take its place. This would be a time of anomaly or crisis in the development of that area of science, as we will discuss below when we consider Thomas Kuhn's views.

Also, the term *theory* is usually reserved for those well-confirmed hypotheses that are considered significant: that is, they are thought to be important by scientists because of the range of observations covered by the hypothesis.

Notice how differently scientists use the word theory as compared to how it is used in common speech. When someone who is not a scientist wants to suggest that something is speculative or uncertain or untested, they will often say it is "just a theory." That can lead to confusion. What most people call a theory, scientists would call a hypothesis. At the same time, because science is dynamic and does not possess any body of knowledge that is considered absolutely true, proven, or correct beyond any possible doubt, the uncertainty indicated by the word theory is at least somewhat appropriate.

An example of a hypothesis that was not totally true but was an important step in scientific understanding and progress in genetics is the "one-gene one-enzyme" hypothesis that is credited to Beadle and Tatum.[5] The hypothesis was formulated in the early 1940s to express the idea that every gene in living things is coded for an enzyme. Beadle and Tatum were awarded the Nobel prize for their formulation and testing of this hypothesis. Since that time we have come to realize that the one-gene-one-enzyme hypothesis was an oversimplification in several ways. For one thing, making an enzyme can require the information from two or more genes: that is, the products of several genes can combine to form one enzyme. Also, some genes do not code for enzymes. Does that mean that the prize should be taken back from these two? Not if one understands how science works. The work of Beadle and Tatum was an important step toward understanding how genetics works. While their hypothesis has been modified, it has not been totally discarded. It represented the best approximation available to them on the basis of experiments done at that time. It was a significant step forward in our knowledge. Later experiments have allowed us to refine the hypothesis as our knowledge has grown. That is a normal occurrence in science.

A few very important and well-confirmed theories have come to be referred to as principles, or laws. Even these can be subject to refinement, as Newton's laws of motion were by Einstein, but refinement is not refutation. With principles and laws our confidence that we are close to the truth is very high, close to an accurate description of some aspect of nature.

There are several places in our list of steps in the scientific method where creative thought is required. This is especially true in the formulation of hypotheses and in the design of tests of hypotheses. Being a good scientist requires imagination and creative insight governed by reality testing. There is no set algorithm for making hypotheses or designing experiments in science.

II. What Science Is Not

You should take note of what the practice of science is *not*. As you can see from the steps listed above, it is never blindly believing what an expert or authority tells you. Just because someone has said that such and such is true does not make it so, and does not lend strength to the hypothesis. Only if the authority or expert has done research or knows of the results of research to test the hypothesis in question does his or her expertise count for anything. There was a time, hundreds of years ago, when the authority was asked and the answer was believed. Indeed, there was a time when, if one expressed disbelief or a different view, one could be put to death for it.

In ancient literature one can find arguments that really look strange today. Some of these came from the deeply seated belief that humans and the earth were at the center of all existence, and that everything in the universe was made for humans. Looking up at the stars at night, and realizing how small and relatively insignificant humans and the earth itself are in terms of size in the universe, it is hard to imagine how we ever could have thought ourselves to be the center of everything. As it is, we sit out in one spiral arm of the Milky Way galaxy, near a sun that is one of a hundred billion stars in the Galaxy, itself one among a hundred billion galaxies in the universe. Even today, there are some who have strong religious beliefs that they attempt to pass off as science. We will deal with one example a little later.

Science is not the gathering of "scientific facts." Notice that I have avoided using the word fact up to now. The term fact can be useful, but to some, facts sound like things that are totally correct and unchallengeable. The knowledge of science is not like that. As I discussed above, even the seemingly simplest of "facts" about the volume of a solution in a pipette involves assumptions.

The term "scientific truth" also overstates what can be known with certainty. Science may progress toward the truth, but any claim to have arrived would be premature. One may find the terms fact and truth used occasionally by scientists, but most scientists are more cautious than to describe their discoveries as so solid as to be immune to doubt. So the term fact should be always be taken with cautious reservation, and use of the word truth in reference to a scientific observation or conclusion is probably best avoided altogether.

At the same time, we should not go overboard with modesty about what we know and don't know. For instance, the earth was once thought to be flat, but a growing body of evidence pointed to its having a round shape. That evidence has grown to such an extent that astronauts now can *see* that the earth is round, and no reasonable person can believe that it is actually flat. We might debate about what kind of "round" the earth is (spherical, slightly pear-shaped, or whatever), but our knowledge of the earth being a round, as opposed to a flat, body is not going to change.

Obviously one cannot always be doubting everything or we would have no foundation on which to build a science. The pure skeptic does not make a good scientist. To do anything worthwhile in science today one makes assumptions, but one needs to be prepared to challenge the assumptions and question them. It is a balancing act, and one's fortunes as a scientist often depend upon getting the balance right. Some theories are so well confirmed that any correction of them is likely to be small, at the most. An example of a small correction with significant implications would be Einstein's correction of Newton's theories of motion.[6] Newton's theories were changed by Einstein, but his theories of motion were never totally discarded. Indeed, Newton's formulas still underpin the entire engineering profession. In many situations, Newton's approximations still do a wonderful, and highly accurate, job. However, if one considers objects moving near the speed of light, the errors in Newton's theories become major. Einstein's corrections led to a fundamental rethinking of the nature of space and time.

Science is also not believing everything one reads in a scientific journal. The process of publication goes like this. A scientist submits a paper for publication. Then the editor of the journal sends the paper out for review by other scientists. The reviewers often demand modification in the paper or more experiments. If a scientist has done an experiment, and claimed a certain result, it is in the nature of science for others to question the assumptions, methods, reasoning, and conclusions. Even the best of scientists can make mistakes or overlook problems in experimental design or interpretation. In their papers, scientists report their methods so that others can repeat their experiments. Results must be repeatable by others, and probably will be repeated by others if the claims are at all significant.

Today, papers in scientific journals are the most common place to find and follow the development of scientific knowledge. Indeed, journal articles and books have been the main place for reporting the results of scientific experiments for hundreds of years. The only way to get and keep current in a scientific discipline is to read the original literature. Perhaps more scientific information will be exchanged on the Internet in the future, but I suspect that the same process of review of scientific material will continue.

III. Mistakes, Disagreements, and Resolution in Science

Whenever even the simplest experiment is performed, there is a chance for error, and when experiments are complicated, as many are in science today, it becomes very hard for one scientist to duplicate another's experiment exactly. Sometimes scientists, or groups of scientists, will come to opposite conclusions after performing an experiment. One or both may have made an error. How does science deal with such a situation? In different ways. Sometimes the two scientists or groups will get together and try to do the experiment jointly. Sometimes other scientists will attempt to replicate the experiment. If the results are of only minor importance, perhaps nothing will be done about it for many years, if ever. Sometimes, too, results of other, different, experiments eventually clarify the situation.

It is not hard to see how errors can be made, and that reinforces the need to repeat experiments. Even something as simple as measuring the freezing point of water can go wrong. There might be contaminants in the water, which may reduce the freezing point. One's thermometer might not be accurate. One might make the measurement too late, after the water has fully frozen and dropped below the actual freezing point. Real science is no different, and scientists should be cautious about their claims because of the risk of such errors.

Historians of science seem to dwell on incidents where a disagreement exists and impinges on the work of a number of scientists. Social issues can come into play in the short run, but there also are self-correcting mechanisms that appear to operate in the long run. There is nothing about the scientific community that makes it infallible. It is possible for a near consensus to be reached on the wrong side of an issue, but it appears unlikely that such an error could continue indefinitely. Especially if the hypothesis or theory were significant and affected the design of other experiments, the fact that a wrong turn had been made would become obvious as further data are gathered.

IV. Other Models of Science

There are other models of how science is done. Some of these are on a larger scale than the description of the scientific method given above. Others are alternative descriptions of some, but not all, scientific studies. What follows is far from an exhaustive list but will highlight a few significant alternatives. These can sometimes give one ideas about how best to formulate and test hypotheses.

One of these alternatives comes from Hegel's dialectic, which describes intellectual advances as coming from the posing of a *thesis* (we would say "hypothesis") and its *antithesis* ("conflicting hypothesis"), with a resulting *synthesis* ("synthesizing hypothesis"). The synthesis is a merging of the concepts in the thesis and its antithesis. Not all of science develops in this way by a long shot, but there have been developments in science that seem to follow this sequence. I'll give two examples.

One concerns the rhythms which most organisms possess. The continuing contraction and relaxation of the heart muscle, with a repeat interval (period) in humans of about one second, is such a rhythm. Another is the human female menstrual cycle, which probably once was once governed by the lunar cycle.

A number of other rhythms repeat about once every day, and these are called *circadian* (Latin: *circa*, "about" and *dies*, "day"). Our body temperatures, our blood-sugar levels, and our levels of activity all undergo changes as a function of the time of day. Early circadian-rhythm research developed around two conflicting hypotheses: the thesis that circadian rhythms were externally imposed by cyclical events in the environment, such as the day-night cycle, and its antithesis that circadian rhythms were internally generated. There was a resulting synthesis: that there are internal clocks that "run" circadian rhythms, but these clocks can be entrained (in terms of period and phase of the rhythm) by environmental events, such as day-night cycling. This synthesis is the accepted view today of how circadian rhythms work.

A second example concerns a debate that raged in physics for several hundred years: The thesis was that light is a wave. Its antithesis was that light is a particle. The ultimate synthesis was that light has a dual—wave and particle—nature. This synthesis is a bit oversimplified here, but gives the general idea. Technically, waves and particles are defined as mutually exclusive categories. Observations have shown that light, however, does not fit those definitions, but appears to have qualities that reflect a combination of wavelike and particle-like properties. Quantum mechanics describes the result.

A productive side of this thesis-antithesis-synthesis view of some developments in science is this: when you hear two persons strongly arguing over conflicting hypotheses, take a little time to consider whether both sides might be partly right, then try to design an experiment to test the possibility.

Another model of how best to do science grew out of the heyday of molecular biology in the 1950s and 1960s.[7] Namely:

Devise conflicting hypotheses (A and B and C . . .).
Do critical experiment (A or B or C . . .).
Recycle the procedure.

Thus, two or more alternative, conflicting hypotheses are formulated as an explanation of a phenomenon. Then one designs (preferably) a single experiment whose outcome should clearly distinguish among the conflicting hypotheses—one will be confirmed while the others will be rejected by the results of the experiment. Perform the experiment, and bingo, one has rapidly progressed toward a single, correct hypothesis.

This is lovely when it works, and it is always worth doing one's best to develop conflicting hypotheses and to design a critical experiment that will distinguish among them. Unfortunately, it is not always possible or easy to do. Also, I've found that after designing what one thinks is the critical experiment to get a yes-or-no answer, Mother Nature often responds with a "maybe" result.

One clear example of this kind of science at work is the Meselson-Stahl experiment on the semiconservative duplication of DNA molecules by living organisms. In a single experiment Meselson and Stahl were able to test three different hypotheses concerning how the genetic material DNA is duplicated. DNA consists of two long strands wrapped around each other in a helix. Those two strands are very precisely duplicated, and the information in the DNA serves as the basis of heredity. Meselson and Stahl asked whether the two strands of DNA, as a copy is made, stay together (conservative replication), come apart, with each strand gaining a new partner (semiconservative replication), or come apart into little pieces (dispersive replication). They were able to design a single experiment that gave a clear-cut answer: replication of DNA is semiconservative.

A longer-term overview of how science is done comes from Thomas Kuhn, and is expressed in his book *The Structure of Scientific Revolutions*. In this book he takes the perspective that science and scientists are, at least to some extent, bound and limited by the conventions of their times. What I have referred to as axioms and assumptions get a more global redefinition by Kuhn, who refers to packages of such axioms and working assumptions as *paradigms*. These paradigms are shared sets of working assumptions and guides to what is important. Kuhn describes scientific revolutions as the overthrow of one paradigm and its replacement by another.

Kuhn defines "normal science" as the kind of science that is done every day by scientists working within a paradigm. During the course of such efforts, occasionally, an anomaly will occur, that is, there will be an unexpected experimental result that does not "fit" the paradigm. I would add that often such anomalies are found to be poor experiments or the result of poor data analysis, but occasionally the anomaly is real, and the result really does not fit the paradigm of the day. The anomaly might lead to other experiments which produce new anomalies, and this can lead to a "crisis," as scientists begin to lose faith in the existing paradigm. That sets the stage for a "revolution," which consists of a new paradigm replacing the old one. This model of how science shifts from one paradigm, or world view, to another has come to dominate the thinking of most historians of science.

The fields of the sociology of science and the history of science are themselves young, developing areas. As they become more scientific, they may come to contribute more to our knowledge of science and scientific thinking. It is a healthy development. However, some who are studying the sociology of science have extended and overexaggerated Kuhnian views to the point that they claim that science is totally relative, that there is no justification for choosing one paradigm over another, and that there is no progress in science. Such views are quite silly, at least from my perspective. There is good reason to choose Einstein's paradigm over Newton's, and the reason is in the data that support the one over the other. We will return to this issue in a later chapter. If you are interested in these issues, I suggest Susan Haack's *Manifesto of a Passionate Moderate* as a good antidote to those who might either deify or demonize science.

V. Science, Nonscience, and Nonsense

A number of years ago I attended a couple of sessions at a convention of historians. What really surprised me was the strength of their assertions where the data were so limited and the issues and variables so complex. In established areas of scientific research, one does not commonly find this kind of insistence that one's hypothesis is right unless the hypothesis has a great deal of experimental support—many tests and many confirmations). Most (though not all) scientists tend to be conservative about what they assert, and recognize that even a single error can have long-lasting impact on one's standing and reputation. In the scientific literature one finds frequent use of such words and phrases as "might be," "suggests that," "one possibility is," "perhaps," "it may be that," "it is suspected that," "this finding is consistent with," and so on. In fact, scientists often go overboard with double hesitations and reservations, such as "it is *suspected* that such-and-such *might* be the case."

My example from one history convention is in no way intended as a slight on historians generally. Theirs is a very complex social science, so complex that some will not admit to history being a field within science at all. I believe that a scientific analysis of some aspects of history is quite possible, but such an analysis will always be very incomplete and much less certain in the strength of its observations and hypotheses than the physical sciences. For one thing, we cannot perform controlled experiments; indeed, we can only analyze and gather the data (or make the observations) that happen to be available. For human history there is the additional problem of not being able to re-create events. One cannot go back and test whether World War II would still occur after changing a single variable, such as not having Hitler at the head of the German government.

But strong assertions in the face of limited data are not limited to historians. Even in those sciences that are founded on firmer and more complete foundations one can find the same kinds of assertions in the less mature areas of those sciences. One example might be the ongoing debates relating to the origin of life. There are two major hypotheses, which can be succinctly characterized as "protein first" and "nucleic acid (DNA or RNA) first." At present, we know relatively little in a definitive way about the actual process by which life originated on earth, so either view, that proteins evolved first or that nucleic acids were the forerunner of proteins, could be correct. But the debates are very heated. It is almost as if scientists in newer areas of science, or areas where data are scantier, feel the need to defend their views with greater emotion. Perhaps with the higher level of certainty in a developed, more established area of science comes the confidence that allows one to express concerns and reservations more openly. Kuhn has discussed the very different ways of doing science that characterize new areas during the so-called preparadigm stage of early development of a field.

One clear message in all of this is that individual scientists are not always logical. We have the same human limitations as everyone else. Fortunately, weaknesses in the thinking of one scientist can be offset by the strengths in that of others.

VI. Certainty of Scientific Results

Let us take a look at the history of one concept, both to see how ideas can develop and to see how the certainty with which we hold hypotheses and theories can grow. Consider the history of attempts to understand what matter is all about. In the fifth century B.C. Empedocles held that there were four different kinds of "stuff"—earth, air, fire, and water. One might take three of these as the equivalent of our present-day solids, liquids, and gases. Then things got more sophisticated as chemists began to identify fundamental elements. They came to realize that such elements could combine to form more complex substances. Lavoisier had identified twenty-three elements by the end of the 1780s, but things really got organized in 1869, with sixty elements then known, when Mendeleyev proposed the periodic table of the elements. From the "holes" in his table, he was able to predict the existence of a number of elements and described something of their expected properties. His predictions were correct. He even predicted that there were errors in the then-accepted weights for a few of the known elements, as he shifted their position in the table according to their known properties, and time has shown him to be right about these as well. It took some guts to do this, especially with the limited knowledge of the day. Mendeleyev had to assume that the order he was seeing would be filled in by new discoveries. Confidence in his proposal by other scientists grew as new results confirmed his hypotheses.

But during the twentieth century we gained an even deeper insight into why the elements have repeating chemical properties, determining the reasons for the differences among elements. In the decade following 1910 scientists determined the structures of atoms, with their protons, neutrons, and electrons. We now know that each element has a specific number of protons in its nucleus: carbon has six, nitrogen has seven, and oxygen has eight. This reductionist insight gives us added confidence that we understand elements and that the periodic table gives us an accurate picture of the nature of matter.

That deeper confidence also allows us to rule out certain things.[8] There will never be an element found that fits between, say, nitrogen and oxygen in the periodic table. That is a certainty because there is no possible number of protons between seven and eight. Because of this insight about the relationship between elements and the number of protons they have, we can be confident that no one is going to announce tomorrow that he or she has found a new element, which we had overlooked, that fits in the middle of the periodic table. There could be more added to the end of the table as we artificially generate new elements. (These heavier elements cannot be found just lying around on earth because such atoms are extremely unstable, and their lifetimes are very short.) We will be getting back to this issue of how good our knowledge of some things is, and I wanted to give Zack a target to shoot at, if he wishes. Some things, I assert, we do know with very high certainty. I believe that the accuracy of a number of scientific theories is as certain, or as near to certain, as human knowledge can be. No reasonable doubt exists for such ideas. But there are many other hypotheses and theories in science about which we are much less confident.

If you are having trouble accepting my example, perhaps owing to a lack of knowledge about atoms and elements, consider the following simpler example. What would you think if someone declared that he or she had found a new continent, one that had been there all along, but we hadn't noticed? Would this be believable? Consider how large a continent is (a new island might actually be possible, especially as these can arise from volcanic action). Consider all of the travels by boat and plane that people have made around the world. Consider the views of earth that astronauts and satellites have had. If someone, say a psychic, announced that another continent existed, would you want to spend your life looking for it? Aren't some things known with at least near certainty?

This has been a rather lengthy tour through the methods of science, but I felt that such an extended introduction was necessary for understanding what follows in our book. More sparks should fly as we move on to consider the human sciences.

Bowen

Science and humanistic literature have a long and not incompatible history. In Plato's day, early humanists—the philosophers and sophists— included thought based on observation and natural phenomena in their catalog of human speculation. But as "science," or early observation, came increasingly to be at odds with religious dogma, it fell into disrepute, owing to a politically powerful religious belief system anxious to protect its own agenda. During the Renaissance, when Greco-Roman secular thought was revived, the absolutism of Roman Catholic religious theology began a decline in the Western world, abetted by the religious civil wars of the Reformation and Counter-Reformation, as the interaction between the humanities and the newly differentiated physical sciences still sought common ground.[9]

Today modern science is on an ever-increasing ascendancy, enjoying all the political, monetary, and reverential assets that have attended its rise. But powerful as its foundations are, science is, after all, a human construct, a branch of human knowledge to be welcomed into the pantheon of ideas. And indeed at least two-thirds of this book represents an effort to define a commonality that we see existing between science and humanistic research. This does not mean that as the former seeks truth through observation and experimentation, it has a monopoly on that indefinable essence, or that it can avoid the hubris and the claims of absolutism of which religion was formerly guilty when it dominated the thought of the Western world.

David very aptly cites the models of scientific research developed in Thomas Kuhn's *Structure of Scientific Revolutions,* based on an evolution of paradigms, or "universally recognized scientific achievements that for a time provide model problems and solutions to a community of practitioners." The term *paradigm* is as old as the Greeks, stemming from the idea "to set up as an example," or in terms of humanistic thinking, to create a narrative that exemplifies a new or unique way of thinking about a specific problem or set of circumstances. Kuhn's paradigm provides a new frame of reference for subsequent investigation and "a change in the perception and evaluation of familiar data" and, should the idea prove feasible, an evolution/revolution in scientific thinking. From my perspective, the flexibility of Kuhn's paradigms is important in describing a dynamic evolving science. But its very flexibility links it to the nonscientific world, even as it comes closer to providing satisfactory answers to universal understanding.

Kuhn later explains that paradigmatic changes require an elitist group to ratify their validity:

> The group that shares them may not, however, be drawn at random from society as a whole, but is rather the well-defined community of the scientist's professional compeers. One of the strongest, if still unwritten, rules of scientific life is the prohibition of appeals to heads of state or to the populace at large in matters scientific. Recognition of the existence of a uniquely competent professional group and acceptance of its role as the exclusive arbiter of professional achievement has further implications. The group's members, as individuals and by virtue of their shared training and experience, must be seen as the sole possessors of the rules of the game or of some equivalent basis for unequivocal judgments. To doubt that they shared some such basis for evaluations would be to admit the

existence of incompatible standards of scientific achievement. That admission would inevitably raise the question whether truth in the sciences can be one. . . .The scientific community is a supremely efficient instrument for maximizing the number and precision of the problem [sic] solved through paradigm change.[10]

Kuhn's is indeed a convincing argument for a belief system that cannot be questioned by outsiders without uniform enculturation, training, and credentials. And it creates a monopoly on what many scientists (not including Kuhn), when pressed on the issue, would declare to be truth. Efficiency demands the creation of a hierarchy to adjudicate the validity of any new paradigm, and that act involves empowerment of a select group to rule on its validity. That empowerment is precisely what we will discuss in a subsequent examination of Foucault's Power/Knowledge.

An additional problem is that Kuhn's argument is essentially a narrative of scientific history and, as a quasi-humanistic pursuit, cannot be verified in the traditional scientific manner. As Jean-François Lyotard tells us in The Postmodern Condition,

The scientist questions the validity of narrative statements and concludes that they are never subject to argumentation or proof. He classifies them as belonging to a different mentality: savage, primitive, underdeveloped, backward, alienated, composed of opinions, customs, authority, prejudice, ignorance, ideology. Narratives are fables, myths, legends, fit only for women and children.[11]

This is not to discount the built-in checks that David describes, checks such as control of all conditions of the experiment, so that it can be replicated at will under similar conditions. These carry the freight of an artificial re-creation of natural circumstance by control over everything except the variable element of the experiment, the only way reliable scientific data can be obtained. The process epitomizes the impulse to exercise control—power—which resided formerly with God, as scientists attempt to wrest knowledge from him (or her) and spread it around the closed scientific community. It sounds suspiciously like the Old Testament narrative tradition. All the more so because of science's adherence to Laws like the remainder of those tablets Mel Brooks brought down from the Mount.

David, perhaps because of my heckling him about scientific verbiage, uses the concept of the "Laws" of nature only sparingly in his description of scientific methodology, and encases even that in admonitions about the flexibility of scientific thinking and the readiness of science to accept ideas contrary to its body of basic legislation. Such humility is not the stock in trade of other textbook introductions to science (including the one chosen for collateral reading for our course, Trefil and Hazen's *The Sciences: An Integrated Approach*), where Newton's Laws of Gravity and Motion and the Laws of Thermodynamics, and so on, are regarded with the same holy respect as the U.S. Constitution, which certainly may be amended, though with great difficulty, but never wholly ignored or defied. It stands to reason that once canonized, the Laws become the foundations on which the whole of scientific credibility rests, and from which the quintessentially expanding constructions of scientific progress draw their logical basis and sanctions.

You have to go through six historical definitions of *law* in *Webster's Seventh New Collegiate Dictionary* before you get to anything that has to do with natural events. The first six definitions carry the burden of being binding, or "enforced by a controlling authority." In adapting such a legalistic concept to nature, scientists consciously or subliminally have proclaimed their own belief system inviolable, and subject only to modification by their fellows in the scientific parliament who enacted the legislation in the first place. Because I am as much a supplicant and believer as anyone else at the altar rail of science does not mean that I do not resent somewhere in my irrational mind the rejection of any inherently creative idea, simply because it violates scientific statutes. And I am not alone: the enormous *X-Files* audience out there are devoted to cults as much because of their violations of "Laws" as because they fervently believe in the supernatural or other off-the-wall phenomena. The individual impulse to think creatively is an impulse to freedom.

David's picture of the overall scientific discipline is a reasonable and orderly one, proceeding along fairly rigid methodological lines. Experimental failures become in his view only a little less beneficial than successes—all scientific experimentation becoming worthwhile in ruling out what doesn't work as well as what does. By the same token fiction seeks to pose possibilities that may or may not have a place in the reality of human experience, but the process resents any arbitrary curtailment of what might be.

The inhibition of science, its very rigor in replicating the inventive successful experiment over and over under controlled conditions, relegates everything to a cause-and-effect process culminating in scientific truths, as much the result of the power to control as to observe. Conversely, in everyday human life, the plethora of causes and effects are so various that no human reaction can be predicted or duplicated with certainty from one minute to the next. Thus, intuitive reasoning and ideas are the stock in trade of the humanities, where, to borrow an old chestnut from Pope, "The proper study of mankind is man."

It is a lot easier to define the basic methodological tenets of science because of their very lock-step cause-and-effect sequentiality, proceeding from such relative inviolables as tables of elements and Newton's law of gravitation in linear progression to other problems, whereas literature seeks variation and complexity as its meat and substance. The human imagination is the point of departure for a far wider scope of activities and beliefs, among them the application of a literary imagination to scientific principles themselves, something we try to do in this book. Examples of literary applications of scientific thinking are everywhere and hardly limited to the examples we will offer here. Our perception of possibilities is not to be limited to those in the present texts, any more than to any of the multitude of approaches to the conflation of science and literature now in print. None of these even come close to exhausting the wealth of ideas on the subject. But at this early stage in our colloquy how could I begin to explain the methodology of humanists as if there were any single such entity? True that each of the humanistic disciplines has its own approach to human concerns, and we in literature seem to have our finger in every pie—art, aesthetics, philosophy, religion, even music, and more recently, politics, anthropology, psychology, and the biology of gender difference—in short, the whole hodgepodge of unregulated interests and attractions of human experience. If we have any "Laws" they are intuitive and perhaps, from a scientific perspective, unprofitable for a linear march to the truth. However, they are the open-ended parameters of a discourse that recognizes few boundaries in its search for knowledge of the human condition and free (in theory at least) from such self-imposed constraints as the scientific community has adopted.

In science, Laws and ascending orders of truth provide a rock-bottom basis for a linear progression to knowledge, even though quantum leaps (both literal and metaphorical) occur from time to time; and as demonstrable accomplishment occurs, so respect in the form of money and power accelerates progress on every scientific front. But what do the quixotic humanists ever accomplish? What purpose do their meandering specializations and prognostications ever serve? I suppose one might answer: A remarkably diverse chronicle of human endeavor and intrinsic beauty, a meaning for life that cannot be produced in a test tube.

Perhaps what I've been saying here is an attempt to make a virtue of my inability to define humanistic endeavor and instead, ask the reader to let intuition rather than a chronology of progress define the enterprise. But no matter how great the ideological gulf, both science and humanistic studies proceed from beings of the same species seeking knowledge, and I hope that by the end of the book we will realize that we all have a lot more in common than we have differences in the ways we seek answers.

Can There Be a Science of Humans?

Wilson

We have explored the nature of science and the scientific method with a focus on the natural sciences. There are other sciences that can loosely be called human sciences—psychology, sociology, anthropology, economics, and history. Are these true sciences? How do they relate to the natural sciences? Can there be a science of humans, or are humans too complicated? Does free will prevent a human science from being developed?

Are the social sciences true sciences? Certainly there are things that can be learned about humans, and even about some of the special aspects of humans, by applying the scientific method. However, my best guess is that there will be less certainty about the findings of many of those aspects that make us human. In part, this is because humans and human societies are so complex and, because of this, unpredictable. I will later go into some of this complexity and unpredictability as we consider chaos and the complexity of human brains and behavior. There also will be less certainty about the findings because there are many experiments that one cannot perform on humans—for instance, from a moral standpoint one cannot just go in and cut out part of the brain from a living human just to do a scientific experiment.

Especially in the social sciences, the design and execution of controlled experiments, where one variable is changed and others are held constant, is very difficult to achieve in practice. The ability to do such experiments in physics and chemistry has contributed greatly to the advancement of those sciences. In the social sciences, there often are many variables, and the relations among them are complex. Truly general laws may be hard to develop given the complexity of human behavior and the difficulties of design and execution of experiments.

Of course, there is no hard break between the social and natural sciences. There are aspects of biology that are complex enough and difficult enough that the certainty of hypotheses and theories will be less, at least into the foreseeable future. There also are areas where results directly impact society, and thus tend to come with greater risks of unintentional bias. Historically, unintentional bias has been present in various sciences, and there is no reason to believe that we are immune, although we have devised procedures that reduce the chance of bias, such as double-blind studies.

At the same time, there are aspects of the human sciences that can be developed with a certainty that at least approaches that of other natural sciences. There is a biological basis to psychology, and neuroscientists are exploring that basis as they make linkages between mental functions and what is happening in the brain.

Just as biology has a foundation in chemistry and physics, so the human sciences have a foundation in biology. Perhaps nowhere is this more evident than in the biomedical sciences, which naturally build from biology. We study other animals, and the results often are applicable to humans as well. We study the cells, tissues, and organs of humans and other animals, and the results help us to understand human diseases and disorders. We have high confidence that we understand the genetic basis of sickle-cell disease, for instance, even though that still has not allowed us to cure the disorder.

This can be extended to the view that aspects of sociology have an underpinning in biology. The field of sociobiology, first developed by E. O. Wilson,[1] is dedicated to this view. Because of the complexity of human social interactions, it is too early to tell how successful such an approach will be as a full foundation for sociology, but some insights are inevitable.

Among the social sciences, psychology is perhaps closest to the natural sciences. (Some might make a case for physical anthropology instead.) Shortly I will consider one particular example of an attempt to be scientific about a topic within the field of psychology, intelligence, and look at the history of this particular endeavor.

Can there be a science of humans? Do scientific laws apply to humans, or are humans somehow above such laws? The scientific world view I have presented makes the claim that humans are animals. We are a part of the natural world, having evolved through natural selection; there is no evidence supporting the view that we were some special creation. We exist in the world and are a part of it. Thus far, the working assumption that our actions are governed by the same laws of physics and chemistry that govern the rest of the universe seems to be paying off. If this working assumption continues to hold, it seems to be quite possible that there can indeed be a science of humans.

Evolution and Natural Selection

We can, in particular, examine the role of evolution through natural selection in giving rise to humans, and this examination will allow us to build a little background for our discussion of *The French Lieutenant's Woman*. Science makes the claim that humans have evolved from ancestral nonhuman primates through natural selection. This has profound implications about who we are and where we came from, so we should be sure to understand what is meant by the claim. How do evolution and natural selection work? How have organisms evolved; how have they gotten more "fit" relative to their environments; and what does fitness mean?

There are a number of influences on the course of evolution. The environment, and changes in the environment, certainly influence evolution as they shape what organisms can exist in a given area. Other factors that can influence evolution include genetic drift (the random changes in gene frequencies that occur in smaller populations), and the isolation of populations (as on islands or mountain tops).

Of all the known influences on the course of evolution, only one has a direction to it. Natural selection contributes to populations of organisms becoming more "fit," or adapted to their environments. Assuming that there is genetic variation in a population, that there are limited resources for the population, and that the population is capable of growing sufficiently in number so as to put a stress on those resources, then natural selection will occur. Under natural selection, some individuals, because of their genetic makeup, will be more likely to contribute more offspring to the next generation than those with differing genetic makeups. The genes of these individuals will become more widespread in the population in succeeding generations. In this way, favorable gene variations are selected.

Of course, what counts as being "fit" depends upon the environment. Change the environment, and the genes that will be favored can change as well. It is in the count of successful offspring that Darwinian "fitness" has its ultimate meaning. Thus, natural selection does not necessarily select for the strongest, and not necessarily for the longest-lived, but does select for the organisms leaving more offspring, which themselves also leave more offspring, and so on. Fitness, in this sense, is not physical fitness, but relates to offspring numbers and vigor. That's Darwin in a nutshell.[2]

Natural selection can act as a designer of sophisticated mechanisms even though it relies, ultimately, on random mutations to do its work. These random mutations are the source of genetic variations, and natural selection works by selecting from among such variations. Most of the mutations will be harmful and detract from survival or reproduction for the individual possessing them. However, given a particular environment, a particular mutation might improve the fitness of an individual. For example, the mutation might alter a protein, allowing the organism better to detect red light. If this ability helps the organism survive longer, and so have more offspring, the mutation will become more common through succeeding generations. Such an increased ability to detect red light might lead, through other favorable mutations, to color vision for the organism. Thus, some of the more impressive mechanisms that we possess probably evolved through gradual improvements. In fact, we now know that vision is an ability that has evolved independently several times during life's history on earth.

Looking just at the end products, organisms and their elegantly functioning systems, from vision in animals to photosynthesis in plants, suggests to some the need for a separate designer, but Darwin's theory of natural selection has given us an explanation that eliminates the need.[3]

I can use three questions to probe your understanding of evolution and natural selection. 1. Did humans evolve from chimps? Answer: no. Humans and chimps have a common ancestor, who lived a few million years ago. From that ancestor, our two species evolved. Neither chimps nor humans were around at the time of the divergence.

2. Has evolution progressed from single-celled organisms at one end, to humans at the other? Answer: no. Evolution can be viewed as a growing tree. There are many branches, and twigs on the branches. Humans are at the tip of one of the twigs, but there are millions of other such tips, each of which, were they able, might make the same claims of superiority as we are tempted to make. Some earlier branches and tips have died—

the species no longer are with us, but other tips contain even single-celled organisms, such as bacteria, which continue to evolve just as humans do. Other branches contain plants, and others, fungi. Near to us on our vertebrate branch are other primates, with other mammals a bit more distant, and with birds, reptiles, amphibians, and fish each more distant still.

3. Are humans at the peak, or at some kind of end-point for evolution? Answer: no. Humans are the best thinkers on earth, but that is not the only measure of evolutionary success. Consider our thoughtful invention of atomic and hydrogen bombs—our abilities can have destructive as well as constructive uses. Thinking alone will not necessarily forever give humans the edge we now enjoy over so many other species with respect to controlling our environment. Hopefully we will develop the wisdom to match the knowledge that our thinking has gotten us.

If we just count numbers, humans, at six billion, are not nearly as numerous as ants or roaches. Furthermore, the roaches have been around much longer than we, and have adjusted perhaps all too well to the changes we have made on the earth.

If we just look at the length of time that particular species have been on earth, there are many other animals who have been around, relatively unchanged, much longer than we have. Turtles have been here a thousand times as long, and horseshoe crabs even longer.

If you still are not convinced that we cannot so easily claim to be the pinnacle of evolution, consider bacteria. Their kind have been on earth for several billion years, while our kind, if we count close primate relatives, have been here less than 6 million years. Bacteria are, relatively speaking, simple creatures. In fact, they can be considered to have taken simplicity to an art form, which allows them, under favorable conditions, to double in number in less than half an hour. A generation takes us ten to twenty years. Bacteria also fit a variety of environments very well. In fact, we each have in our intestines more bacteria than there are humans on earth.

So, it is not so simple to pick a single "winner" of evolution. Organisms have evolved to fit their environmental niches, and, while we can adjust ourselves to an range of environments, there are many species that fit as well into their niches as we do into ours. We can't climb trees as well as our close relatives, and we cannot survive in hot springs, as thermal bacteria can.

None of this is meant to diminish the importance of being human.

Certainly we are the most intelligent creature on earth. Certainly there is something special about our ability to contemplate ourselves and our place in the universe. That, after all, is what this book is about, ultimately. But I hope all this has given you some insight into what natural selection is all about.

All evidence points to natural selection as the mechanism underlying humans having evolved on earth, and that leaves us with some distant cousins that make for strange bedfellows. All life on earth is related, as can be seen in our common genetic code. This deep insight into human origins cannot help but influence our personal views of human nature, and perhaps that is why it is only with some difficulty that this now very well-established theory continues to meet objection from some religious fundamentalists.

How Close Are We to a Science of Humans?

But I have digressed somewhat from the question of whether there can be a full science of humans. To some extent there already is part of one. I just was discussing it: humans have evolved. There certainly also is a rather mature science of human physiology—we understand much of what goes on when we take in oxygen and give off carbon dioxide, for example. We know about the genetics of humans and how close our genetic information is to that of other species, such as chimps. We know the genetic cause of a number of human disorders, from Huntington's disease to cystic fibrosis. We know about the physiology and biochemistry of nerve regeneration, of muscle contraction, of hormone action, and so on. I don't think these are very controversial. Thus, a biomedical science of humans has been developing, and few would deny that we have made substantial progress. The progress is visible in our improved health and ability to understand, treat, prevent, or reduce the severity of, a number of human disorders, from smallpox to heart disease.

But some would argue that this is the "easy" part of human science. What about the special aspects that make us human—our minds, our thoughts, our culture? Can we know what thinking and consciousness are? Can we predict whether the stock market will go up or down this year? The real challenges for a science of humans seem to arise when we deal with those aspects of humans that make us special. How scientific can sociology become? In a thousand years, will it be as complete a science as physics is today? It's too early for answers to such questions to be given, but clearly it will not be an easy task, and I suspect that these human sciences will carry with them considerable uncertainties that will remain into the foreseeable future. One of my reasons for such a belief involves the complexity of the human brain, but I will not go into that now. Instead, I will give one example—the ongoing attempt to establish a scientific foundation about something that seems to set humans above other living organisms, namely, intelligence.

IQ Tests and Intelligence

Defining Intelligence

Our first problem is in defining what we mean by intelligence. Is it just one thing? Some who were involved in the development of so-called intelligence tests believed that it was possible to have a single scale, or intelligence quotient (IQ), which would indicate each individual's general problem-solving ability. In a kind of circular reasoning, IQ was taken to be whatever intelligence tests had in common.

For many of us who have taken the different parts of an SAT test or GRE exam, there seems to be an oversimplification in attempting to measure human intelligence with a single number. Certainly some of us score much better on the math portion and others on the verbal portion of these tests. Then there are other kinds of intelligence to consider: musical, mechanical, and so on.

Consider intelligence reflected in manual skills. Or compare the kinds of thinking and problem solving done with language to those that involve spatial thinking, as in map reading or taking apart or putting together something like a bicycle or auto engine. Have you ever tried to read and follow written instructions for building something of even moderate complexity? Sometimes the words seem only to make the job more confusing. Then you look at a picture or drawing and everything comes clear. Some of us do better at visual exercises than with language; others cannot make sense of the relationships among mechanical parts. It would seem that there are different kinds of intelligence, and some of us are better at some kinds than at others.

So, part of the science of intelligence will be to get a handle on these different kinds of abilities and the relationships among them. That knowledge gathering probably will not be complete until we have mapped many of the brain activities involved in different kinds of problem solving. But, placing this large difficulty aside, we still can take the oversimplified concept of IQ to measure something. What gives us that IQ? Is it genetically built in, or are there strong environmental components to it?

Genetics of IQ

There are apparent differences among individuals in terms of intelligence. There also are apparent differences among individuals in terms of IQ scores. Intelligence is one of those characteristics that are influenced by a number of different genes, and also by the environment. In this context, "environment" should be taken broadly to include what happens during embryonic and fetal development, early childhood experiences and nutrition, and so on. But some have argued that one's IQ score is largely fixed by one's genes—that most of the differences in IQ among individuals is the result of heredity. This becomes an important issue to the extent that IQ can predict, in part, one's success in life in terms of education, career, socioeconomic class, and so on. Here we have a complex issue that needs to be untangled, and the history of this debate certainly shows some of the weaknesses that characterize the early development of human sciences. Oversimplification seems to be a mark of many of the weaknesses and shortcomings. It also has been (is still?) difficult not to carry a bias into these debates on genetics and IQ, and that is not a productive way to carry out a scientific analysis.

Is the difference in IQ mostly in the genes? The history of this debate is as old as the concept of IQ, but we can pick it up in the late 1960s with Arthur Jensen's monograph.[4] Jensen reviewed a number of studies of the heritability of intelligence within populations, and concluded that up to 80 percent of the variation in IQ scores seen among individuals in a population was genetic. The publication of Jensen's work produced an explosion of heated debate. For example, William Shockley, a physicist and apparently a racist, picked up on Jensen's results to argue that blacks, who averaged lower than whites on such tests, were intellectually inferior to whites due to genetic differences. Just as things seemed to be calming down, Herrnstein and Murray published their book, *The Bell Curve*, which has ignited the same issues and discussion almost as if they hadn't already been debated twenty years earlier.

To understand the limitations of the debate, it might be useful to examine what a heritability of 80 percent actually means. It does not mean that 80 percent of intelligence is genetic, or inherited. That would be a misreading on several levels. For one, what is being looked at with heritability are differences in intelligence among individuals, not overall intelligence. For another, heritability is not the same as inheritance. Heritability measures, for a given population, the fraction of variation (which is related to the differences in IQ scores among individuals) accounted for by genetic differences among individuals. Any such measure of heritability holds only for that population, with the environments that the individuals experienced. Here, "environments" means all of the influences on a developing individual, beginning with fertilization, embryonic development, childhood, puberty, and into adulthood. And, if there is any change in the environments, and range of environments that individuals experience, heritability can change. Thus, heritability is not a fixed number, but depends upon genetic and environmental ranges that are present or experienced by a population. Heritability is a function of the range of environments available—the smaller the range of environments, the higher heritability.

In the United States, the average black scores about 15 points lower on IQ tests than the average white. Is that difference due to heredity—to genes? Environments for the average black and the average white in the United States certainly are different. Such environmental differences can greatly influence heritability measures. One indication that IQ differences might be more environmental than genetic comes from the observation that SAT scores are a poorer predictor of success for African Americans than for whites, for whom its predictive value still is limited. Of course, many human characteristics important to success in life are not measured by IQ—motivation, perseverance, leadership ability, and interest. And, as was discussed above, IQ is only a limited measure of the complex character we call intelligence.

There also is the possibility of racial, and other, bias in the questions on IQ exams. One way to see that such bias can exist is to examine a few questions from IQ tests that were given long enough ago that cultural biases are more apparent. Lewis Terman, one of the early developers of IQ tests, gave sample IQ test questions in his 1916 book, *The Measurement of Intelligence.*

A question for eight-year-olds was: "What's the thing for you to do when you have broken something which belongs to some one else?" (215). Terman states that "Satisfactory responses are those suggesting either restitution or apology, or both. Confession is not satisfactory unless accompanied by apology" (215–16). Of course, the acceptable answer appears to have more to do with moral values than with intelligence.

One question for fourteen-year-olds was: "There are three main differences between a president and a king. What are they?" (313). Terman states that "The three differences relate to power, tenure, and manner of accession. Only these three differences are considered correct" (313). "Unsatisfactory contrasts [include] 'A king wears a crown.' . . . 'No differences, it's just names.' . . . 'A president lets the lawyers make the laws.' . . . 'Everybody works for a king'" (314). The ability to answer this particular question seems much more an issue of knowledge than intelligence.

Another question was: "A man who was walking in the woods near a city stopped suddenly, very much frightened, and then ran to the nearest policeman, saying that he had just seen hanging from the limb of a tree a . . . a what?" (315). According to Terman, "The only correct answer . . . is 'A man who had hung himself'" (316). Terman goes on to say that "There is an endless variety of failures: 'A snake,' 'A tiger,' 'A cat,' etc." (316). Someone who grew up in a tropical environment might well have a different idea as to what is an acceptable answer to the question. Furthermore, Terman does not remark on the possibility that others had done the hanging of the individual, a very real possibility at the time he was writing. But perhaps the reaction of being frightened and running to a policeman was intended to rule out that possibility!

Terman's adult IQ test includes the "Problem regarding the path of a cannon ball" (333). The subject is asked to draw the path of a cannon ball that is fired from a horizontally positioned cannon. The only satisfactory solution has the ball beginning almost on a level and then dropping more rapidly toward the end of its course. Terman adds that "Any one who has ever thrown stones should have the data for such an approximate solution" (334). Gaining the knowledge of the trajectory the cannon ball would follow is not as simple as Terman suggests. He may not have realized that Aristotle had quite a different opinion about how objects moved through the air, and I find it unlikely that Aristotle had not thrown some stones during his lifetime.

The entire field of racial differences and intelligence has a sadly biased history. It is reviewed in a skillful manner by Steven J. Gould in his book,

The Mismeasure of Man. Gould reviews the last couple of hundred years of the (so-called) science of human brain measurement and intelligence. He points out the systematic errors that were made. I'll give one example. In the middle 1800s it was popular to measure the size of human skulls, to give an idea of the size of the brain, and it was presumed, without any good data, that the larger the brain, the more intelligent the individual. Caucasian males were doing all of the measuring. They collected skulls from around the world, and guess what—they concluded that Caucasian males had the largest brain sizes. When the measurement and analysis errors are corrected, today, we find that there are some differences in average skull size in this particular set of skulls. Of course much of this difference tracks overall body size. But when the errors are corrected, some of the skulls from nonwhites exhibit brain sizes as large as those of Caucasians. In the earlier studies, the skull sizes matched the expectations of the time for who should be most intelligent. In the end, there is not much evidence that brain size alone, within humans, reflects or limits intelligence.

Obviously, this example of skull-size measurement shows the presence and risk of bias in science. We will return to the question of bias later in the book. Here, it can be said to be something to watch for, especially in the human sciences. Ingrained cultural expectations can mislead even those attempting to be neutral and bias-free.

Getting back to IQ, if heritability of intelligence were high, and were intelligence to be unchangeable by any reasonable environmental manipulation, one might have some reason for concern that we were creating a locked system, where individuals have great difficulty rising in social and economic status, whether they be black or white. But we see migration among the classes as individuals go through life, and as one generation gives way to the next, despite the strong social forces that retard such interclass migration. We will consider a locked system from science fiction later as we examine Huxley's *Brave New World*.

One telling illustration of the weakness of the hypothesis that things are locked, that one's IQ is fixed by genetics, and that the poor will always be poor, comes from an analysis of average performance in the United States on intelligence tests administered between 1918 and 1989. As first noticed by John Flynn,[5] raw IQ scores actually increased dramatically during the twentieth century, not only in United States but in a number of other developed countries. Relative to the average score of 100 in 1989, someone in 1918 would have scored only 76 on a similarly scaled IQ test. Are those who took the test in 1989 really that much more intelligent? Doubtful. Some who took the test at earlier times are still alive, and they do not seem to lack in intelligence compared to those who are younger. Then what gives here? Answer: we don't even know for sure! And that ought to give us great pause in thinking that we know what today's score differences among individuals really mean. Certainly the difference between the average score in 1918 and that in 1989 is not genetic because there have been too few generations to show that kind of difference in IQ due to genetic changes. So, we can conclude that dramatic changes in IQ scores are often linked to environmental changes, even if heritability is high, and even if we don't know exactly what the relevant changes are between the environments of 1918 and 1989.

What about average SAT scores, which, everyone is noticing, declined during the latter part of the twentieth century? There are two possible explanations I can think of: 1. More are now taking the SAT today, and the additional individuals are not, on the average, as bright as the select few who took the exam thirty years earlier. 2. The SAT is not just an IQ measure, but measures other kinds of things as well, such as specialized knowledge.

Today, we do not know whether there are racial differences in average IQ that are caused by genetics. We do not even know, assuming there are such differences, which races would be "on top"—black would be as likely as white, with Asian somewhere—above, below, or in the middle. Why this uncertainty? For one thing, we do not experience the same environments, on the average. The average black in the United States occupies a lower socioeconomic status than the average white, and that environmental difference contributes to lower IQ scores. We also still have some bias and prejudice out there. The environmental differences can begin before birth, in utero, and that early developmental time can be critical to later intellectual development.

Second, it is not at all clear that the cultural definitions of race have anything much to do with genetic differences among humans anyway. For instance, in the United States, anyone with one black great-grandparent, or even one great-great-grandparent, the others being white, has been called black, whereas genetically such individuals are much closer to Caucasian. In addition, only certain external features, such as skin color, are used in society to make such judgments. Genetics goes well beyond skin color. And the recently completed human genome project as well as other data suggest that our classification of individuals into races does not account for a majority of their genetic variation. There is more variation within each race than between races. I also remind you of what I said earlier about the likely origin of humans, out of Africa, and the consequent greater genetic variation among native Africans than among whites. The most likely scenario for how humans arose and spread around the globe indicates that all *Homo sapiens* arose in Africa, and spread from one part of Africa to the rest of the world beginning about 100,000 years ago. This is somewhat oversimplified, since there is evidence of gene flow both from and back to Africa (at least, from Asia), so there was some intermingling, probably after the "out-of-Africa" event. Nevertheless, there is more genetic diversity among humans in Africa than there is in the rest of the world, which dispels any simple view of IQ being lower among blacks because of genetics, since blacks are more genetically diverse than the rest of us. One might argue that among blacks there would more likely be a wider range of genetic potential for intelligence, both higher and lower, but even that is a speculative view.

Third, given this nation's emphasis on individuals, even were there to be average differences in intelligence between groups, the only possible issue would be at an individual level, and there is much variation within each group. Indeed, even with today's environmental differences, there is more variation of IQ scores within each group, Caucasian, Black, and Asian, than there is between groups. So perhaps we should forget about the misleading issues of race and just consider the existing differences in intelligence among individual humans more generally.

Fourth, it can be misleading to try to dissect out the genetic and environmental components of a complex character like intelligence. The mix of many genes that influence intelligence and the complex interplay between those genes and environment during individual development certainly point to risks of oversimplification if one tries to separate genetic or environmental components or contributions to an individual's intelligence.[6]

Where Do We Go from Here?

There are differences in intelligence among individuals, and there are resulting differences in how quickly different individuals learn, and in what kind of learning works best for different individuals. Some of us learn faster than others, for whatever reason, genes or environment and experience. Some of us learn better if we hear something, others if we see something. Probably our educational systems should be designed to recognize and adjust for such differences; usually they aren't. We have individuals who are not learning the basics in school, whether because of heredity or environment. Do we give someone who isn't learning enough an extra year or two to learn, or do we argue that doing this would be bad for their morale, and so just continue to promote them, and then graduate individuals from high school who can't read?

The question of the existence of genetic differences in intelligence among individuals has become very political. Some people seem to be especially frightened about the possibility. That concern might be reduced if people would realize that genetic differences do not mean unfixable differences. Let me give a pair of examples that relate to skin color. Let's compare people with dark skin and people with light skin. There is a genetic component to skin color, with several different genes involved. Persons with darker skin have a real advantage over those of us with lighter skin: exposure to ultraviolet (UV) light causes less damage to those with darker skin. Darker skin protects from sunburns and skin cancers by blocking out such UV light before it reaches deeper layers of the skin.

So why did pale skin develop among some who migrated out of Africa 100,000 years ago? Perhaps because they migrated to places where less sunlight reaches the earth. Away from the equator, there is less need to be protected against UV light because there is less of it. In addition, in such regions, there is an advantage to having light skin because the body makes use of some of the UV light to generate vitamin D in the skin. That generation is greatly reduced if UV light penetration is blocked by darker skin. A little winter sun on a pale face in Scandinavia will generate sufficient vitamin D, but a dark-skinned child is more likely to get rickets in such northern latitudes. So, in such environments, there was good reason for lighter skin to evolve. But now we palefaces are developing an increased risk of skin cancers, especially when we visit the beach or move back toward the equator, like going to South Florida to live, and even more so with the thinning of the ozone layer.

Notice that I've described two problems here: lack of vitamin D for dark-skinned individuals, and sunburns and skin cancer in those with lighter skins. Both these problems are related to the genetics of the individuals, but there are easy environmental adjustments that allow us to "fix" both of these genetically influenced problems. The lack of vitamin D is easily overcome by adding vitamin D to diets, and that is what we have done with vitamin D–enriched milk. The risk of sunburns and skin cancer can be reduced by the regular wearing of sunscreen. Thus, genetic differences that handicap particular individuals in particular environments can be correctable. We should not be so fearful of genetic differences in intelligence—it is much better to work with them, and to overcome them, than to deny them.

We can dump on the oversimplified concept of IQ, and can dump on possible biases in particular tests that are used to measure it. But we should not mislead ourselves about our talents or weaknesses; we should face up to, and work on, our individual limitations. We should not make differences, when they are a handicap, more of a handicap than they need be.

Returning to our more general themes, I hope you can see how we can make a scientific approach to a subject near and dear to humans, namely, intelligence. The limitations on such an enterprise are obvious. In this case, the subject is complex, measurement is difficult, and preconceived notions and bias have crept into many of the attempts to develop such a science. If this example is at all typical, it appears that it might take generations to get the science of humans even onto the right track, and an analysis of intelligence is just a little piece of the puzzle. But these human sciences are quite young, and I don't think that we yet know how scientific our understanding of humans, and especially those aspects that contribute to our uniqueness, can become.

There is another issue, however, that we need to address, one that is at the heart of the question of the nature of humans and the possibility of developing a science of humans. That is the concept of free will, and a discussion of it is coming soon.

Bowen

I am daunted by David's description of the orderliness of science's formulaic lockstep approach to truth, and my inability to offer anything even resembling that for literary studies, let alone the humanities as a whole. Humanistic studies, of course, focus primarily on human beings, and, apart from the arts, generally confine their study of inanimate objects to their relationship to people. When modern science was in its infancy, during the late eighteenth and early nineteenth century, both science and humanities embraced the study of nature, reenforced by what was called positivism:

> At its core positivism was an assertion of faith in the method of observing nature in order to infer regular, law-like patterns of phenomena. The phenomena which presented themselves to the senses of the scientist constituted reality for positivists, who were skeptical about metaphysics of any kind and hence about conventional religion. Positivism was a movement which, although firmly rooted in experimental science, had far-reaching social and political implications. Its reformist rationalism valued science as the major tool for social progress. . . . Thus, while the actual practice of science was becoming more and more focused and specialized, claims about the universality of scientific method were being extended.[7]

Many humanists and scientists were not particularly devout, and tended to treat the observation of natural phenomena as far more reliable than religious belief. When science became more rigidly structured in its methodology, the gap between the two major disciplinary areas widened, a schism abetted by the reorganization and departmentalization of universities.

The humanities had been attempting to understand the human species and our environment long before science hardened its methodological arteries. If the principal difference in methodological approach forms the most apparent contrast between sciences and humanities, the second division is probably the way we approach our understanding of human beings. Science is content to discern a place for mankind in a scheme of natural phenomena, while the humanities are more interested in emphasizing the study of human beings per se.

As the interests of the two disciplinary areas converge in genetics, particle theory, and human nervous systems, and we begin to view human beings as simply another species increasingly subjected to the scrutiny of the scientific legal system (laws, theories, phenotypes, and the rest), this seems to some humanists to demean what they have been about for a couple of millennia. Before we get too indignant, we should recognize that scientists are human beings too, people with coherent answers to age-old human questions, even if some of the answers scientists have provided often exponentially increase the threats to our survival and complicate our understanding of the purpose of human progress.

History, the narrative of past events, is perhaps the oldest of our formal disciplines, still as active and dynamic an area as it ever was. The positioning of history in the formal roster of university department affiliations varies in larger institutions. It has been classified severally with the humanities and with the social sciences, and, as we shall see, shares some of its empirical methodology with the sciences; so it provides a good transition to our social sciences discussion.

Historians today approach their subject matter with far different agendas and methodologies than did their predecessors, as revisionists constantly hold previous notions of what and why events occurred up to newly evolving points of view. When David, attending a history conference, was appalled at the emotional conviction attached to speculations unsubstantiated by experimental verification, he was applying a scientific standard of controlled conditions that could never be used to determine the meaning of human behavior. But, as he admitted, one can't recreate World War II under controlled laboratory conditions to verify a hypothesis. This does not invalidate the need for speculative interpretation of history, of course, but merely asserts that scientific validation is hardly a prerequisite of human knowledge, or that the only kind of useful knowledge has to be sanctioned by scientists. In a way that is one of our problems.

Science has had a long and sharply contested struggle to replace religion as our primary belief system, and it is only natural (to use a controversial metaphor) for science, like religion, to insist on purity of purpose and methodology. But with the enormous progress of scientific discovery came the concomitant conversion of a large segment of the population to belief in a secular order of both thinking and the rehierarchicalization of methodology and creed. We can only hope that the blessings that ensue will be accompanied by accommodation, recognition, and respect for a variety of disciplinary and philosophical approaches to human endeavor. That is a major reason why David and I are writing this book.

While history embodies traditional cause-and-effect reasoning, it creates its own semifictional narratives, taking whatever fragments of documentation are available and applying them to hypotheses emanating from the historian's cultural conditioning. I am not so sure that the narratives of science operate much differently, since the methodology admits only ideas that fall within the strictest parameters of what constitutes proof as defined by scientists. That certainly is cultural conditioning.

For now it suffices to say that one can't always conflate the infinite varieties of human behavior with present scientific observations of natural phenomena in the exterior world. Historical research is the quintessential example of cause-and-effect methodology. Like anthropology or archeology it uses records, verbal and written accounts, and artifacts to create new, often dialogically opposed, narratives to counter old assumptions embodied in former narratives. It does not seem odd to me that, like scientists, historians believe in the truth of their results, at least until new artifacts and narratives come along to alter their assumptions.

Now that David and I are momentarily on the safer middle ground of the "soft," or social, sciences, we are freer to cast disparaging eyes on the "rigor" of their treatment of scientific methodology on the one hand, and their lack of sensitivity to the individual and unique on the other hand. In effect these hybrid science/humanities disciplines came into being in part because of the fairly recent (seventeenth century) popular success of scientific methodology, in order to apply it to the activities of human beings.

I agree that the social sciences have experienced some mixed results—but still more than enough to warrant some respect, money, and power in public political arenas. Its practitioners appear as television "talking heads" after all cataclysms, natural and social disasters, and more recently after most public child-abuse and domestic arguments. An army of social counselers is summoned to the vicinity of every disaster to console the victims and survivors and to augment or replace the diminishing role of the religious priesthood in that task. That social scientists are not to be wholly trusted by politicians is evidenced by their promotion of social schemes designed to correct certain patterns of destructive behavior when more primitive methods popular with a conservative public might suffice (executing murderers, allowing the poor to starve and be deprived of medical treatment), those time-honored measures of bygone revenge patterns and natural selection.

While the claim of the social sciences to authentic truth is that they are indeed scientists, their justification in the eyes of "hard" scientists like David can hardly be fully credible simply because social scientists can't exercise complete control over human beings or their environment (except in cases like the use of test victims in Auschwitz or Tuskeegee). Still the behavioralists do use the fundamental principle of detailed, systematic observation, much as nineteenth-century naturalists did. They use mathematical computation to justify and support their conclusions, even though in examples like the ones David cited involving intelligence tests, they sometimes ratify preconceived notions, no matter how far off base.

And their biases may not be limited to the familiar trope of skin color. For instance, a whole set of "primitive" parameters was developed regarding the Irish to justify British imperialism in Ireland. That the Irish were more closely related to apes than the English was ratified by anthropological studies indicating smaller Irish brains, sloping foreheads and bent stances. All these attempts to malign were published in both scientific journals and popular magazines like *Punch*. Similarly unflattering patterns of stereotypical behavior were systematically cited in evidence by social scientists, declaring "discoveries" involving colonized populations throughout the world.

This behavior raises a more fundamental question: Were these defined characteristics the products of prejudiced selection, or is every observation characterized by the preconceptions of the observer? That possibility is one we will come back to in our later discussion of the relationship between reality and our perception of it. For the moment, it is enough to say that human understanding depends as much on conditioned preconception as on innocent observation. We fit what we see into the master narratives of our lives, and these are shaped by our previous assimilation of every influence we have encountered in the past.

This negative characterization of occasional past abuses by individual social scientists is more than counterbalanced by the enormous contributions of the social sciences to our understanding of the human condition, and society's attempt to alleviate its problems. These disciplines, given grudging respect because of their quasi identification with science, are in many ways our principal hope of social progress, of seeking authoritative, creative, practical solutions to our most pressing social problems. Exactly what those problems are is perhaps our biggest conundrum, and not one easily answered by that great sociological tool, the public opinion poll, in which truth is inadvertently linked with what some mathematically significant proportion of people conjecture at any given moment.

That brings us to the subject of psychology. David's discussion of psychology reflects an appreciation of only that aspect of the discipline that is most congenial with scientific belief in physical causality for even the most complex and as yet not fully understood aspects of human behavior. The treatment of aberrant behavior has become associated more recently with mind-bending drugs, rather than psychotherapy. It began with the observation that frontal lobotomies (the destruction of part of the brain) seemed to quiet unruly people, as did electric shock treatment. With this knowledge of mechanical physical solutions to mental activities and with the advent of the particle theory of energy, neuroscientists could at last come to grips with human thinking, and materially influence the process through physical means. Thus that segment of psychology became a branch of science in itself, since it represented a form of control that the more amorphous and unwieldy psychotherapy lacked.

On the other hand, many of us in the humanities embraced the other more mystical aspects of the observations of Sigmund Freud, a practitioner whose experiences with his patients and his own personal psychopathology produced observational narratives, immune to experimentation, controls, or quantification. Few people on either side doubt that Freud was trying to be honest, but fewer scientists give him the credit he deserves for his discovery of a mind divided into a conscious and an unconscious. Humanists, particularly those involved in literature saw the potential meaning of this bifurcated idea of reality which accounted for so much distress in a society programmed to assume that learned behavior, no matter how constricting, was the only reality.

It is fashionable now for experimental psychologists (seemingly the majority) as well as many literary critics to marginalize Freud. His descendants, however—Jung, who extended the subconscious to include tribal behaviors from time immemorial, and Lacan, who formed even more exotic narrative scenarios of pervasive subliminal behavior—found adherents in a variety of ideologies. The influence of these men's ideas on our everyday life, on advertising and commercials, and on the explanation and manipulation of modern cultural behavior in general is enormous. It creates a healthy skepticism toward every received truth. For instance, whenever I read or listen to any sort of political analysis, I want to know what its source is, who or what institute or think tank sponsored it, printed it, and paid for it, and how the material presented is related to their agendas. Freud in a sense has trained us to be critical close readers or observers of the subtext of public announcements, before we can decide what can legitimately be considered true.

Finally, I would like to comment briefly on the political and social aspects that have ensued from another great scientific pronouncement, this time one derived, like Freud, mainly from observation, and only more recently subjected to experimentation, Darwin's *Origin of Species*. Here too was a popular narrative so convincingly written that everyone except biblical fundamentalists could buy into it, and even elaborate on it.[8] If the principle of natural selection could be demonstrated in human reproduction and its relation to environment, it could also illuminate humans' ability to cope with the economic and social environment. Survival of the fittest had a strong affinity for Calvinists and other religious practitioners for whom fitness was directly related to piety. Prospering economically and socially was ascribed to the morality of the upper classes. Other populations, subject to religious and social domination, as well as economic victimization, were subjected to the rule of the strongest, the more righteous, and those genetically suited for elevation. The combination provided ample rationale for justifying means as well as ends in a Hobbesian world that is naturally mean, brutish, and short, and it eventually spawned a capitalist mentality whereby the accumulation of wealth conferred honor and served as the basis for a new nobility, supplementing the old nobility of birth, particularly among our European ancestors. Capitalism, economic Darwinism, and the rewarding of individual initiative metamorphosed into the American dream, currently suffusing the global economy. The promise of perfection through adapting to conditions is more than survival; it is a consummation devoutly to be wished. In rationalizing human greed, economic Darwinism is certainly an unintended perversion of a popular scientific theory, but so was the atomic bomb.

4

Darwinism, and Knowledge and Power

Historical Perspectives and Politics
in Science and the Humanities

Bowen

The Origin of Species is one of the most obvious scientific demonstrations of cause and effect, in that the process it describes rests on a history of the genetic variations of a species as it adapts to the circumstances of its environment. The process of adaptation may be speeded up in the laboratory, but in the main, evolutionary theory retains its validity by tracing the history of such adaptations under natural conditions. As David observed, skin color may genetically adapt itself to varying degrees of exposure to the sun in different climates, whereas other differences are far more subtle. I will not pursue again the numberless variations in human beings or their multifarious environmental conditions, except to say that Darwin's notion has been accepted by most of us in the humanities as a valid point of departure for our own concerns. We have seized on its cause-and-effect paradigm as common to behavioral patterns far beyond the physical milieu. We tend now to interpret even human history and the rise of power in all its forms as variations on Darwin's survival of the fittest. Although I discussed this concept earlier in more general terms, I would like to apply it now to the dichotomous view of human knowledge as science and the humanities, and to whether or not the pursuit of knowledge has anything at all to do with the power structure.

Both groups of disciplines pursue knowledge, using different methodologies, as we have seen. I want to speak now in more theoretical terms about one aspect of the motivations that lie behind work in the academic disciplines and discuss changes that our attitudes have wrought upon the nature of our several quests. To begin, I will call your attention to a lecture by Michel Foucault, delivered on January 7, 1976.[1] The lecture represents one segment of the thinking underlying the work of Foucault, one of the most influential writers in contemporary social criticism. (My interpretation of this text, in isolation from the prolific body of Foucault's work, is a personal one and applies directly only to issues involving the relationship of science to humanistic study as I see them.)

Much of Foucault's work seems fragmentary and disconnected, even though that trait serves to authenticate his pronouncements against control, in a sense becoming his antithesis to the thesis of scientific methodology. He writes of the inhibiting effect of global totalitarian theories, such as Newton's laws, monotheism, and the like, on the free search for knowledge. These overarching theories eventually become tools for sustaining the control by the ruling class, forces that work to rationalize and affirm the strength of that class by subjugating the knowledge of individuals to their controlling agendas. Opposed to these is the "local character of criticism," which creates an "increasing vulnerability to criticism of things, institutions, practices, discourses," through an "autonomous, non-centralized kind of theoretic production, one . . . whose validity is not dependent on the approval of established regimes of thought" for its reality.

Foucault sees a new "return of knowledge" through "subjugated knowledges," or "the historical contents that have been buried and disguised in a functionalist coherence or formal systemisation." These include such examples as the evolving exegesis of conflicting biblical texts to cohere to the beliefs of a hegemonic priesthood, the construction of a coherent national history that gratifies those currently in power, a system of laws that protects the sovereignty of the state, and, for our immediate purposes, a natural order that justifies capitalism. Hence the appropriation of Darwin's survival-of-the-fittest theory, intended to apply only to reproduction of species, as David has told us, evolved into a doctrine that uses the weight of newly acquired scientific reasoning to provide justification for the wealth and power of the rulers.

The discredited or disregarded knowledges exist in many forms—including oral tradition, as well as written documentation—stemming from innumerable sources, individual and collective. The knowledges possessed by such individuals as incarcerated psychiatric patients, disenfranchised illiterates, and colonized "primitive" populations are discarded or disregarded by the ruling hegemony. Foucault foresees an insurrection of combined subjugated knowledges that ultimately concern a "*historical knowledge of struggles.*"

Many of these knowledges are "erudite," leaving behind documentation of one sort or another, documents that repeatedly reflect their struggle for recognition. The "*genealogy,*" or union of erudite knowledges and local memories, "allows us to establish a historical knowledge of struggles and to make use of this knowledge tactically today." The idea has infused contemporary research in the humanities, as a few examples will show. Early research on gender, then called "feminist criticism," found common cause with research on the suppression of racial and religious minorities, diasporic populations, and colonized populations everywhere. Offering documentation such as slave narratives, diaries, plantation account books, on one hand, and medical testimony of eighteenth- and nineteenth-century "rest cures" for self-assertive wives, as well as accounts of foot binding, and so on, on the other, subjugated knowledges have brought genealogy into the current discourse. The history of outrages perpetrated against Native Americans and Caribbean native populations by Western European colonizers, until recently overlooked, is still debated, but a new awareness of what the hegemony has swept under the rug has given rise to the recent reluctant approval of "cultural studies" in the nation's schools, even as the countertendency toward "privatization" is suspected of sheltering children from any but the approved canon of study.

"But what," you might well ask, "does any of this have to do with science?" Simply that the discourse of science has increasingly come to frame the pursuit of knowledge. Examined through Foucault's power paradigm, the sciences are clearly winning the day and accruing prestige and resulting power within the academic community and with the general public, creating a measure of antagonism that inhibits our joint pursuit of human knowledge. Beyond that there lies the temptation for science to become more and more vulnerable to appropriation by the prevailing power structure, on one hand, and its own more rigorous exclusionary tactics on the other. Foucault poses a question for today's scientific community:

What types of knowledge do you want to disqualify in the very instant of your demand: 'Is it a science'? Which speaking, discoursing subjects—which subjects of experience and knowledge—do you want to 'diminish' when you say: 'I who conduct this discourse am conducting a scientific discourse, and I am a scientist'? Which theoretical-political *avant-garde* do you want to enthrone in order to isolate it from all the discontinuous forms of knowledge that circulate about it?

A prime example of this sort of exclusionary approach to "scientific truth" is embodied in Rothman's *The Science Gap*, which we examine in detail later (see appendix A). The creation of a genealogy is, then, an attempt to emancipate historical knowledges "in opposition to the scientific hierarchisation of knowledges and the effects intrinsic to their power."

As was said before, the investigation of people rather than other natural phenomena poses such problems that the "hard" sciences (chemistry and physics) will validate research involving human beings only if it directly depends on scientific laws, rules, and regulations in its experimental procedures. Thus biologists, like my friend David, are well within the scientific community if they apply the laws of physics and energy to their investigations. Modern medicine is for the most part the product of such reasoning. Quasi-scientific investigation, such as psychoanalysis (as opposed to work with mind-altering drugs) is suspect once it strays from the foolproof methodology of scientific verification. Thus, the "soft sciences" of sociology and cultural anthropology are often more suspect than accepted in the elitism of science. They of course are dealing with enormously complex human reactions, and the honored simple cause-and-effect shibboleths of science do not easily apply. That is where the wild tribe of humanists often find common cause. Not only are humanists primarily interested in people, but their observations nearly always take the form of narratives—at best, for scientists, narratives of probability rather than the fictive narratives of possibility (of what might be) that infuse literature.

This is not to say that scientists never tolerate a spoonful of narrative to make their experiments go down, but when the narrative, plausible or not, becomes fictive speculation, it is dismissed to the la-la land of literature. But such exclusionary thinking can inhibit cross-fertilization among disciplines. Let me offer an example: Now that literary scholars are increasingly finding common cause with anthropologists in contemporary theory, significant segments of each group are working in close alliance in the pursuit of human knowledge, and often their metaphors come from hard science itself. Kath Weston's "The Virtual Anthropologist," in attempting to explain the postmodern concept of postcolonial hybridity (the mixing of cultural influences between the conquered and the conquerors), finds an apt metaphor in chemistry: The hybridity factor is like the combining of two chemical agents—compounding produces a new unique substance, while mixture permits the substances to remain intact.[2] This example is only one of innumerable possibilities open to those with a smattering of interdisciplinary understanding. The study of science and literature is in itself one of the fastest-growing areas in literary theory. Other literary works, such as those we will discuss in this book, are infused with and informed by science in a way that exploits the possibilities of both, though at present the obverse (literary studies contributing to scientific research) is only occasionally and incidentally true. That was not always the case. For example, the ancient science of astronomy is permeated with classical allusion—the planets and stars have the names of classical deities, indicating the alliance between ancient sciences and mythology. But now it is rare to see scientific terminology borrowed from literature.

Even as scientists struggle to preserve their purity, occasionally subjugated knowledges, like acupuncture, may be brought into the scientific fold if and when they have withstood the traditional experimental tests, but full acceptance is conditioned on appropriation of their discourses into those of the scientific community, just as scientific discourse itself was appropriated by a power structure in the process of demoting religion, pure reason, and intuition.

What does all this mean in practical terms for the sciences and the humanities? While education, especially higher education, has long enjoyed a love-hate relationship with the political/economic power structure, more and more interdependence and cooptation are the price some scientists must pay for an alliance that has only marginal relation to truth. The hegemony hardly thrives on dissent and conflict, and in its attempt to resurrect subjugated knowledges, the humanities has lost a lot of the favor it may have found in bygone eras. Each year as the battle is newly joined with recently recovered revisionist ideas and research, the finances provided for the humanities decreases proportionately, while scientific research, ever more costly in the aggregate, looks increasingly to governmental and "private" (read capitalistic) support as scientists shift their generally funded work to practical profit-making or power-accruing (read applicability to weaponry) research, as well as break-throughs in medicine that can be trumpeted nightly on TV and daily on political rostrums. One set of disciplines may now be seen as querulous, uncooperative radicals, while the other is providing humanity's salvation.

Of course this jaundiced representation may be seen as totally biased griping, and it is certainly not applicable in every case; but it does ring true, at least partially, within a research university where "research" often means funded scientific research from which the institution derives an inordinate amount of its income. Nor should it be inferred from my skewed analysis that scientists don't often come up with results or opinions contrary to those that dominate capitalistic/political dogma. When the government goes looking for evidence that the world is not cooling, or that the elimination of whole species by economic fiat is merely a minor matter, they are often hard-pressed to find scientists outside corporate-owned think tanks willing to sacrifice their professional reputations for political propaganda.

On the other hand, the now subjugated knowledges of the humanities are further jeopardized by their own internecine warfare, involving the practitioners of traditional intrinsic aesthetic values and the social critics of our present situation. Most of us in my discipline still find it impossible to discuss literature without bringing something of both views to the discussion, any more than in general education we can disavow the importance of science in the history of the human race. In spite of the competition and intellectual arrogance on each side of every disciplinary divide, there is more than a mere hint of admiration, intellectual curiosity, and the genuine desire to investigate largely unexplored avenues of mutually beneficial cooperation. We hope this book may make some contribution to that effort.

Wilson

As Zack describes Foucault's views, they seem extreme exaggerations to me. As I understand the argument pertaining to science, it is that scientists have "sold out" to the rulers of the day, that everything is about power, and that there is no reason to think that scientific knowledge is better justified than any of Foucault's "subjugated knowledges."

Do some scientists "sell out?" Do some "rulers" make use of scientific advances, and the technology resulting from those advances, to further their own ends, even at the expense of others? Yes on both counts.

But such a reality does not lead to the generalizations and conclusions that Zack describes. Science, and its results, can be used by the strugglers of the world as much as by the rulers. Scientific knowledge can be used for good or evil. Scientists seem as often to be challenging authority as supporting it, and that would be expected if their major goal is to try to get at the truth. Certainly the results of Galileo and Copernicus were blowing against the prevailing political and religious winds of their times. And Zack himself points out that global warming is a problem that scientists have been studying and that many of today's governments have been trying to avoid taking account of. One could add scientific studies of the consequences of air and water pollution, the loss of biodiversity, acid rain, and the risks attending the introduction of exotic species to the list, as well.

Earlier, I pointed to Einstein's modifications of Newton's laws to show that scientific laws do not represent fixed, unchangeable rules, and I will give several other examples in a later chapter on scientific revolutions. Science is modifiable.

"Subjugated knowledges" are another matter. There are a variety of claims and beliefs in the world, some more justified than others. Consider claims about such medical treatments as folk remedies, herbal medicines, and acupuncture. Are these "remedies" effective and safe? How are we to tell? Is Foucault claiming that we can just accept all the claims as true? I certainly would rather have scientific tests done to determine if herbal "remedies" are safe, and to determine whether they actually help any medical condition. For such medicines, it would be useful to purify the active ingredients from the herbs, especially since it is known that the content of such ingredients can vary significantly in individual plants. If there is a safe and effective ingredient, or combination of ingredients, in a given plant, wouldn't it be better to take a known dosage? Submitting the claims to testing seems the only way to know if they actually do work. What else are we to do? Go back to just accepting the claims made by others? Return to the era of snake oil and cure-alls?

In the case of acupuncture, some limited and some wild claims have been made about what kinds of conditions it can effectively treat. A few of the more limited ones may be found to have merit, and studies of acupuncture, sponsored by the National Institutes of Health, for a variety of conditions, are now under way. Now, to some, NIH is an establishment entity, so we have a case of a subjugated knowledge actually being tested, as Zack points out. But for some of the claims that have been made it will probably turn out that acupuncture is no more effective than a placebo, and people who rely on acupuncture alone may be risking the development of more serious disorders that could have been avoided by other existing treatments.

The "subjugated knowledges" of Foucault seem to consist of a hodgepodge of ideas, some of which are not worthy of the term knowledge, but are more like claims or untested hypotheses, and for many of these, the methods of science actually offer a way of separating wheat from chaff. Other "subjugated knowledges" appear to refer to the views of oppressed individuals: for them science, and the technologies that flow from it, can be just as much a boon as it can be for the ruling class.

Scientific testing allows hypotheses and ideas to gain a level of certainty unavailable by mere assertion. I see no program or proposal put forward by Foucault that would allow one to determine the validity of the claims being made by his "subjugated knowledges." Science may not be a perfect mechanism for determining the validity of claims, but it seems far ahead of whatever is in second place, and in the case of Foucault I don't even see a second mechanism being proposed. I do not accept that all statements have equal merit or accuracy—it is too easy to make obviously false claims.

Zack brings up psychoanalysis, and suggests that it is considered unscientific. Indeed, in all of its baggage of unsupported particulars, it probably is, but that does not mean that one cannot test its usefulness as a therapy. Recent results suggest that, for some mental disorders, a combination of drug treatment plus a "talking treatment," such as psychoanalysis, works better than either one alone. However, for other mental conditions, it may turn out that psychoanalysis does not help any more than a sympathetic listener would. One does not just reject psychoanalysis out of hand. Instead, one does tests to see if it is effective. I suspect that we will learn that psychoanalysis is not correct in many of its detailed assertions about what human beings are all about, if for no other reason than that humans are so complex, and the hypotheses of psychoanalysis are too simple to encompass everything. Nevertheless, aspects of talking therapies may indeed help some of us.

Zack also speaks of an alliance between the power elite and scientists.

To the extent that governments and companies are major underwriters of the costs of doing scientific research today, I agree that there is a real risk of distortion of results in some cases. To the extent that such distortions occur, other scientists, less tempted by short-term gains, might well correct the errors. There has been a history of corrections of errors by scientists, no matter what the cause of the errors. Beyond moral arguments, any temptation to cheat on results is offset by the risk that exposure of such cheating will bring ruin to the reputation of the individual. Among scientists who accept government or corporate support, most are not just pushing the ruling elite's line—something else is driving the scientific enterprise. I sincerely believe that most scientists are interested in pursuing the truth, and the great majority of us are more driven by the desire to understand than by the desire to control. I do not believe that the gulf between humanities faculty and science faculty is all that great on many issues.

Clearly some research areas are advancing faster than others because of more funding by governments or corporations. Other possible subjects of scientific research can be limited in their progress, or greatly slowed, by decisions of such governing authorities, at least to the extent that significant dollars are required to carry out the research. Research on fusion power or solar power may go underfunded if oil companies lobby Congress to shift research dollars toward fossil fuels.

In a democracy, if people are unhappy with the decisions of their elected representatives concerning such funding, they can elect others. Of course peoples in democracies also can press for the elimination of research in certain areas, and that actually is a concern. The resulting science will not necessarily be wrong but it will be incomplete.

Sometimes a lack of data in a field can be damaging, as when we lack knowledge of the details about an ecosystem needed to protect an endangered species, or when funding for studies of global warming are cut because some are afraid the results will cost them something. Some prefer that humankind have its head in the sand on certain issues, especially if there are personal benefits that they can continue to enjoy because of our collective ignorance. Restrictions on funding can influence the body of knowledge that we have, but they are not the result of scientists' own decisions, and are frustrating to scientists engaged in studies of ecology, global warming, or solar energy. Such areas of ignorance, while perhaps benefiting a few in the short run, can cost the rest of society in the long run. The solution to this kind of funding problem rests with all of us, not just with scientists. Elected representatives need to ask what is in the best long-term interests of the nation and the world. I know that sounds idealistic.

So, in the end, is it all a question of power and money controlling scientific results and making them invalid? Science seems to have gotten some things close enough to right so that technology works. For instance, imagine a humanities professor, an ardent supporter of Foucault, pulling his dinner from the freezer, and heating it in the microwave. A quick drive in his automobile takes him to the airport, where he boards a jet aircraft. In flight, he starts up his laptop to put the finishing touches on a soon-to-be-delivered speech asserting that scientific claims to knowledge are no stronger than the "subjugated knowledge" claims of astrology or phrenology. What's wrong with this picture?

Bowen

At the risk of protracted bickering, and with all due deference to my friend and colleague, I feel compelled to make a couple of observations on David's response. Having laid down the hard and fast parameters of scientific testing, David asks for a comparable set of proofs for every idea, every subjugated knowledge—proofs that would satisfy scientists. Ideas do not seem to have a life of their own until they are voted on by scientists. Thus scientists set themselves up as arbiters of knowledge, or at least the kind of knowledge in which they have an interest.

This leads us back to the issue of money and power. David conflates two issues: the kinds of research being done, and the integrity of that research. I am happy to concede that there are very few scientists who succumb to falsifying their results to please their sponsors, but I am less willing to gloss over the idea that scientists' plans for research are unaffected by the willingness of grant-giving institutions, both corporate and governmental, to fund specific areas of scientific research rather than others. Academic research institutions promote and nurture grant-getters, while dismissing others. David admits that this has a sometimes deleterious effect on the kind of knowledge produced. David also admits the necessity and reality of governmental and corporate support but ties it exclusively to the rare resulting distortion of experimental findings. The point is that to a large extent scientific research is controlled by money, and money is associated with power and control.

To offer the pious opinion that in a democracy we can always elect the honest and turn out the self-serving is an exercise in naïveté, especially in light of the current debate on campaign financing. We have already made bribery legal, and in the current climate entrenched legislators, beholden to corporate and ideological interests for their mind-numbing propaganda machines, are all a part of a continual power struggle. Certainly the social sciences and humanities are affected, but the predominant stature accrued by science should not be put in the position of falling prey to such pressures. I think David would agree with me on that point.

Finally a comment on David's delightful concluding remarks. What's wrong with his picture is that, as the hypothetical humanities professor, I hate frozen microwaved foods that taste like cellulose; I have never had "a quick drive" to the Miami airport (and neither has David); I am often prohibited to use my laptop for some ill-defined, scientifically generated reason; and I never read papers on astrology or phrenology. So what has science done for me lately?

5

Darwin and
The French Lieutenant's Woman

Bowen

John Fowles's *The French Lieutenant's Woman* re-presents, develops, and expands upon Charles Darwin's *The Origin of Species*. It seems to me a prime example of how fiction can provide new insights on the human condition by drawing upon the accomplishments and narratives of science. The novel is primarily about evolutionary changes that occurred over the hundred-year time span between 1867 and 1967. The narrator/ author uses anachronism by developing the earlier scene in cognizance of the evolving history and ideas of the intervening period, often casually inserting allusions to twentieth-century events and mores in the narrator's descriptions—for example, the observation that "There would have been a place in the Gestapo"[1] for the particularly disagreeable old shrew, Mrs. Poulteney. Yet Fowles assumes the character of a Darwinian paleontologist in examining his characters as fossils of a bygone era, and recreating that era in terms of the "well-made novel," a phenomenon of the late-eighteenth and nineteenth centuries, commonly thought to represent the acme of novelistic writing, with its perfect blend of form, moral action, and structure. The language both of the characters and of the narrator evoke the prose of that era, and the characters bear a striking resemblance to a gallery of nineteenth-century classical character types. For instance, Sarah Woodruff has overtones of Hardy's Tess and Sue Bridehead, Dickens's Miss Wade, and Eliot's Dorthea Brooke; Ernestina sounds a little like Thackeray's Amelia Sedley, and Dickens's Esther Summerson; the wise aunt, Mrs. Trattner, might have come out of *David Copperfield* as David's Aunt Betsy; Mrs. Poulteney bears a resemblance to Dickens's Miss Haversham; and the servants Mary and Sam seem almost Pickwickian.[2] Using such dated features as epigraphs at the beginning of each chapter, Fowles depicts the action of his characters within a setting of Victorian mores, along with poetry, statistics, excerpts from essays, and factual reports published in the nineteenth century, each providing some detail of social, intellectual, or reportorial background against which to view the action of the episode. Thus the reader becomes, as it were, a kind of archeologist of the nineteenth century, as Fowles presents his own lengthy disquisitions on the Victorian temperament, fashions, class structure, and politics.

Set in Lyme Regis, with its fossil-laden shoreline, the book, like the region, is a fertile area of investigation into the past. Charles Smithson, its male protagonist, is a Darwinist by avocation, intent on examining the fossilized artifacts of history, and the reader, like him, is invited to participate in that activity. The town laid out behind the shore line is viewed as if through a Darwinist's glass:

> A picturesque congeries of some dozen or so houses and a small boatyard—in which, arklike on its stocks, sat the thorax of a lugger—huddled at where the Cobb runs back to land. Half a mile to the east lay, across sloping meadows, the thatched and slated roofs of Lyme itself; a town that had its heyday in the Middle Ages and has been declining ever since. To the west somber grey cliffs, known locally as Ware Cleeves, rose steeply from the shingled beach where Monmouth entered upon his idiocy. (10)

Fowles goes on to describe the settlement around the boatyard, focusing in on its open-throated archetypal lugger in terms of an dissected specimen, before giving a thumbnail sketch of the town's evolution from the Middle Ages to Elizabethan times, like the dating of a specimen environment.

Charles is not the only scientific observer, however; the "local spy"— he is later hinted to be Dr. Grogan, himself an ardent Darwinist and medical practitioner—is observing Charles and his fiancée, Ernestina, along with a third figure, Sarah Woodruff, the titular protagonist, through his glass, and speculating on what sort of species he is examining:

> On the other hand he might, focusing his telescope more closely, have suspected that a mutual solitude interested them rather more

than maritime architecture; and he would most certainly have re-
marked that they were people of a very superior taste as regards
their outward appearance. . . .

But where the telescopist would have been at sea himself was with
the other figure on that somber, curving mole. It stood right at the
seawardmost end, apparently leaning against an old cannon barrel
upended as a bollard. Its clothes were black. The wind moved them,
but the figure stood motionless, staring out to sea, more like a living
memorial to the drowned, a figure from myth, than any proper
fragment of the petty provincial day. (10–11)

Through the device of the magnifying lens, Fowles turns the description
of the characters and setting into a gestalt of the historical present, with
its comfortable associations of the past, its acceptance of the 1867 environment, and its wonder at any figure that might constitute an anomaly
to examined species and the normal social ("natural") construct of the
age. Fowles intersperses between the two sections of the above quotation
a description of the atmosphere in terms of what manner of dress was
fashionable, since fashion fads are among the most transient aspects of
society.

Thus the scene is set for Charles's Darwinian fossil collecting along
the shore and for the reader's fossil collecting of the characters and events
of an action dated 1867, a time that reflects the tumults of rapid evolution
and change, and one in which a species of woman, somehow out of her
time, but perennial, might easily be described as "a figure from myth"
who does not fit the Victorian mold, but actually, as we shall see, heralds
the dawn of a new age of women's equality. So different from the ladies of
her time, she is referred to by the observing scientist as "it," a sexless
figure rather than a woman. Sarah stands staring out at the sea from
which she, or her dreams and aspirations, have emerged, cast up by the
desertion of a marooned French lieutenant on the shores of Lyme Regis's
petty provincialism.

Farthest out on the Cobb, a natural jetty, with its salt-encased crusta-
ceans, Sarah is anonymously dressed in plain black, in contrast with the
young couple who are decked out in the fashionable attire and attitudes
of the era. They are dated, she is not. The whole scene is permeated with
references to Darwinian evolution and the Victorian belief that the age
had attained the pinnacle of superiority. Ernestina's father, Mr. Freeman,
a very successful draper, has already told Charles that "Mr. Darwin

should be exhibited in a cage in the zoological gardens. In the monkey house" (12); nevertheless, while not universally accepted, Darwinism, through Charles, Dr. Grogan and other scientifically advanced types, has already gained a substantial foothold. As they walk upon the strand, Ernestina reminds Charles that they are walking on literal as well as literary fossils, thus admitting another major—literary—link between past and present: "'These are the very steps that Jane Austen made Louisa Musgrove fall down in *Persuasion*'" (13). So, in effect the reader joins in the investigation of a literary as well as historical age in the throes of an evolution into modernism and postmodernism.

Currently, we in literature are participating in a heated and to me fairly meaningless debate about what modernism and postmodernism are, and how the two "modernisms" differ. In 1967 Fowles was regarded as a postmodernist by many defenders of the avant-garde not only because he was an existentialist (another barely definable term, which for purpose of this discussion I will call a lack of belief in a manifest destiny), but because in *The French Lieutenant's Woman* he had written a "self-reflexive" novel, or a novel that is in part about the act of writing itself. Authorial intrusions in *The French Lieutenant's Woman* include the notorious chapter 13 in which Fowles debates whether the characters themselves may or may not dictate their own actions and choose their own course of events. And of course Fowles himself appears later as a character in his own book. While literary history has lots of examples of direct authorial participation in the action, Fowles links the outcome of this novel not to conventions of satisfactory closure (the expectation of a satisfying conclusion that will allow everyone to live happily ever after), nor to any of Aristotle's dictates. Instead he seeks in part to provide a book that will proceed to a realistic end, given the circumstances the characters face. And in this sense he comes close to his topic of evolution—an environment that demands adaptation for survival, but yet affords within that parameter at least a hope of freedom of choice. On its surface, Darwinism seems to be an exceptionally deterministic doctrine when translated merely as the survival of the fittest, but, as David has said, there is a lot more to it. Natural selection represents adaptation to the environment, and adaptation implies, at least, the possibilities of choice. In this book, Fowles pretends that the choice for its creator depends on a literal toss of a coin, but the problem is how to make that authorial choice seem the only possible one. But more of that anon.

What Fowles does is overtly inject into his book some of the subjugated knowledges (the discrete but disregarded knowledges of the Victorian period), like the epigraph detailing the disproportionate prewar and postwar ratios of women to men and how the resulting imbalance restricts women's marital/career choices in a life circumscribed by social biases. How is society supposed to regard women like Sarah whose behavior doesn't fit any mold? Are such women part of some pattern of temptation dating back to Eden, or are they simply mentally unbalanced? Remember the discussion of rest cures for assertive wives? How different is Dr. Grogan's suggestion about committing Sarah to his friend's genteel loony bin when she gets under Charles's skin?

As stated previously, Charles and Grogan share a common belief in Darwinism, a passion that might be equated with religious zeal. Under constant attack from fundamentalist Christians, Darwin's work opened new scientific doors to questions about divine creation and theological revisionism in general. Having established their mutual affinity to Darwin, the "two lords of creation" begin their incantation of belief with Charles's assertion "'that Lyell's findings are fraught with a much more than intrinsic importance'" (130). Lyell is described as "the father of modern geology . . . [whose] *Principles of Geology*, published between 1830 and 1833 . . . hurled . . . back . . . [the origins of man] millions [of years]. . . . Genesis is a great lie; but it also is a great poem; and a six-thousand-year-old womb is much warmer than one that stretches for two thousand million" (130–31).

When Grogan produces his new bible, *The Origin of Species*, as evidence of his scientific faith, he affirms its relevance to life rather than death: "'This book is about the living, Smithson. Not the dead'" (131). Later Grogan expands on his homily of text worship by pledging his oath of silence on the text: "he laid his hand, as if swearing on a bible on *The Origin of Species*" (177). He admonishes Charles, "'Man, man, are we not both believers in science? Do we not both hold that truth is the one great principle?'" And then to cap the new religion with its own martyred Christ, Grogan asks, "'What did Socrates die for? A keeping social face? A homage to decorum?'" (179).

Fortified by his religious recommitment to science and Darwinism, Grogan, in his zeal to persuade Charles to have Sarah put away, produces "scientific" testimony in the form of anecdotal documents regarding famous cases of female hysteria. The atrocities of the Roncière case (183–

89) consisted of supposedly false accusations by a teenage girl against an attractive young officer. Convicted on her testimony despite convincing circumstantial evidence to the contrary, Roncière becomes the subject of scientific treatises on female lunacy and the menstrual cycle, bolstered by patient anecdotes offered by a German physician. Following Roncière's retrial and exoneration, Fowles adds, as an asterisked appendage, that information produced subsequently indicated that Roncière was at least guilty of physically violating the girl's person, if not of raping her. Thus Grogan's urging, on infallible scientific grounds, the convenient commitment of Sarah as a melancholic lunatic is apparently nullified by historical evidence to the contrary.

Charles's increasing passion for Sarah begins with Victorian humanitarian, gentlemanly concern for her plight and stress, a concern nurtured by such sentimentalists as Dickens and company, but laid over with subliminal manifestations of lust. Had Smithson had the benefit of Freud or his contemporaries and successors, he might have seen it coming, although it probably would not have mattered much. But since no one in 1867 talked honestly of such things, especially in mixed company, conventions of male behavior as well as polite society represented even greater subversions/perversions of sexuality than occurred during the ensuing hundred-year span.

Ernestina is a paradigm of provincial morality and custom. While she reflects the faults of her upwardly mobile, money-driven merchant class, she is, like her fiancé, a product of her time, and, with all her learned affectations, certainly not a villain or utterly incapable of honest introspection. Her function is to represent the conditions of fossildom, to be a typical specimen of her age.

Charles, the Darwinist fossil hunter, is also a fossil in transition. He can rebel against existing social, ethical, and moral conditioning, but is caught in the web of social constraint:

> He felt that the enormous apparatus rank required a gentleman to erect around himself was like the massive armor that had been the death warrant of so many ancient saurian species. His step slowed at this image of a superseded monster. He actually stopped, poor living fossil, as the brisker and fitter forms of life jostled busily before him like pond amoeba under a microscope, along a small row of shops that he had come upon . . . (230). There was no doubt. He was one of

life's victims, one more ammonite caught in the vast movements of history, stranded now for eternity, a potential turned to a fossil. (262)

Sarah Woodruff, on the other hand, seems like a person out of her time, defying tradition, self-willed, and suffering from the social classification, reduced prospects, and restrictions on women common to her age. Her predicament, prompting a loss of self-esteem and a self-abusive penance, also conversely prompts a stubborn will to survive and improve her condition. In essence she becomes a twentieth-century woman caught in a nineteenth-century environment, and ultimately finds refuge in what may have been the only house in England that had ideas sufficiently advanced and the political/artistic clout to sustain them, that of Dante Gabriel Rossetti, poet, painter, and leader of the Pre-Raphaelite Brotherhood.

As Charles's passion for Sarah grows from pity to solicitude to fascination and then obsession, the course of his love carries the trappings of courtly love evoked in terms of Victorian sentimentality through the poetry of Tennyson, Hardy, Clough, and Arnold, among others quoted in the chapter epigraphs. And even these evoke more antique tales of unfulfilled passion. The story of Charles and Sarah, the sensitive innocent and the mysterious saint/temptress, is as traditional as literature itself. In many ways Charles is a nineteenth-century version of the existential sperm swimming up a postmodern fallopian tube in Barth's "Night-Sea Journey"—an exercise seemingly futile, racked with despair and uncertainty, but "singing . . . 'Love! Love! Love!'" (Barth, 13).

Like his contemporary Barth, Fowles retains a certain modern abhorrence of sentimentality, even while he emulates the nineteenth-century novel form, and, like Barth, mitigates the damages with comic touches. Smithson performs a comic parody of religious repentance in an Exeter church after agonizing over his orgasmic three-second copulation with Sarah. It is a testament to his narcissism that, caught between her lies and truths, and his overwhelming love/desire for her, he rails against the viciousness of his society and sees himself as Christ, perhaps masochistically equating crucifixion with great if brief sex.

He saw himself hanging there . . . not, to be sure, with any of the nobility and universality of Jesus, but crucified.

> And yet not on the cross—on something else. He had thought
> sometimes of Sarah in a way that might suggest he saw himself
> crucified on *her*. (284–85)

Timing is everything in Fowles: eternity in an instant with Sarah, a microcosm of frozen time that recapitulates the species, as it does in the pure survival-equals-reproduction theory of Darwin. What makes Fowles's novel so interesting is that he similarly freezes the infinite continuum of time into a period of two or three years, and, like a modern paleontologist, examines the frame in the whole context of what has transpired since, bringing with it its own inevitable historical, cultural and social conditioning. The reader can hardly fail, even without Fowles's constant reminders of twentieth-century ideas and events, to read this text as a fossilized account that is transformed both by its environment and the biases of the modern audience. Making the book a specimen of the whole Darwinian process, Fowles ingeniously calls to our attention the similar process of recreating an era through fictive interpretation that blends science, documentary evidence, and the creative imagination. In doing so he admits deviating from strictly scientific shibboleths:

> *The Origin of Species* is a triumph of generalization, not specialization; and even if you could prove to me that the latter would have been better for Charles the ungifted scientist, I should still maintain the former was better for Charles the human being. It is not that amateurs can afford to dabble everywhere; they ought to dabble everywhere, and damn the scientific prigs who try to shut them up in some narrow *oubliette*. (45)

As Fowles tells us, Darwin inevitably opened the doors of the imagination to all sorts of variations on the Darwinian scheme of environmental adaptability and survival of the fittest, not merely profligacy in the act of conception, as David has told us, but in social and economic ways. I am sure that David is adhering to the strict scientific point of view in saying that "fitness [in scientific Darwinian terms] is not physical fitness, but relates to offspring numbers." I have no quarrel with that definition, but following Fowles's sentiments expressed above, amateur Darwinists related to the humanities and social sciences have evolved a number of variations on the theme, and scientifically impure but useful narratives such as social and economic Darwinism have taken the holy name and applied it to entirely different, but seemingly analogous purposes. Several of these are demonstrated in Fowles's text.

When Charles's inheritance from his uncle, his ultimate claim to a title and the family fortune, is thwarted by the old man's desire for scientific Darwinian "fitness" by marrying a fecund younger woman and producing at least one male offspring, Charles seizes upon the opening to renounce his engagement to Ernestina on the grounds that his economic circumstances were now so unpromising that he could no longer be a "fit" husband. We have already seen Charles proved scientifically "fit" in the physical sense by impregnating Sarah and producing an offspring with an almost instantaneous ejaculation in a single-thrust coupling with her. Earlier, however, at the point of his discussion with Mr. Freeman, Ernestina's father, regarding his new prospects, Charles is only potentially "fit" in the strict scientific sense. Mr. Freeman, on the other hand, has found a new justification for his amassed haberdashery wealth in the form of proving himself to be one of the "fittest" in terms of economic Darwinism, as he tries to convince Charles to join his firm and earn some of his own money:

> "You will never get me to agree that we are all descended from monkeys. I find that notion blasphemous. But I thought much on some things you said during our little disagreement. I would have you repeat what you said, what was it, about the purpose of the theory of evolution. A species must change . . . ?"
>
> "In order to survive. It must adapt itself to changes in the environment."
>
> "Just so. Now that I can believe. I am twenty years older than you. Moreover, I have spent my life in a situation where if one does not— and very smartly—change oneself to meet the taste of the day, then one does not survive. One goes bankrupt. Times are changing, you know. This is a great age of progress. And progress is like a lively horse. Either one rides it, or it rides one."
>
> Charles did indeed by this time feel like a badly stitched simple napkin, in all ways a victim of evolution. Those old doubts about the futility of his existence were only too easily reawakened. . . . The abstract idea of evolution was entrancing; but its practice seemed as fraught with ostentatious vulgarity as the freshly gilded Corinthian columns that framed the door. (227–29)

Even after disinheritance Charles has enough of an income to retain his status, but not to progress up the evolutionary economic ladder, and he is enmeshed in the net of gentility that prevents his applying himself to mundane trades. So Fowles leaves it up to the reader to decide whether Charles's fossilized excuse for indolence is a blessing or a curse.

The other success story with regard to economic Darwinism is that of Charles's butler, Sam, who seizes on the miscreancy of Charles to feather his own economic nest. Again Sam's action could appear to be in some measure justifiable, since he had suffered under his subordination to Charles, but his actions in any age would seem despicable. In Darwinian terms, however, Sam is partly redeemed by his own natural ability in gauging the taste of the day and by his creativity in assembling an inventive window dressing that attracts customers' attention. He is a born advertising man, and the only one in the book—with the possible exception of Sarah—who achieves his goal. And after that he manages to get off the moral hook and mitigate the damages of his duplicity in failing to deliver Charles's crucial correspondence to Sarah by providing Charles with Sarah's new address.

The other, previously touched on, Darwinian perception is the overt and covert emphasis on time as the necessary driving force. Everything I have discussed so far is overtly related to time, raising still another question: whether time itself is a perennial consideration or one related merely to the hundred-year gap between the events of the story and its narrative relation. In two intertwined scenes, related to the indeterminate future by the two young children—the scene with the prostitute/ mother with an infant in the back room, and the later scene involving Charles and Sarah's daughter—the crucial symbol linking the foreshadowing first scene to the second is Charles's watch, the child-pacifying instrument by which he establishes contact with the future generation. It is also the final object the author/narrator/character, John Fowles, produces to conclude his personal identification with the book:

> He takes out his watch—a Breguet—and selects a small key from a vast number on a second gold chain. He makes a small adjustment to the time. It seems—though unusual in an instrument of the greatest of watchmakers—that he was running a quarter of an hour fast. It is doubly strange, for there is no visible clock by which he could have discovered the error in his own timepiece. But the reason may be guessed. (362)

It is Fowles's excuse to turn back the clock fifteen minutes and provide another, more postmodern, existential ending to the book. The indeterminacy of the ending, mentioned briefly earlier in this chapter, has attracted a great deal of literary-critical attention. Fowles provides three possibilities. The first, thoroughly traditional, ending has Charles, after explaining away his relationship to Sarah, reunited with Ernestina and fathering seven children by her in a respectable Victorian marriage (266); the second has Charles reunited with Sarah and their child, leading another happy life, and coping with the advanced ideas of the new age; and the third has a solitary but wiser Charles rejected by Sarah, but also more able through his experience to cope with whatever destiny awaits him, perhaps, it is hinted, in America. I must admit that on the first reading of the book, I opted for the second ending, being at heart a sentimentalist with a fondness for Victorian fiction. But 1967 would have demanded, even in a fossilized artifact, something more than the snide "thousand violins cloy[ing] very rapidly without percussion" (360) concluding the penultimate closure. Critics have wondered whether, especially in the lines quoted above, Fowles was really serious about providing the freedom of choice he seems to crave for both reader and novelist. Perhaps postmodernism has as little choice as the Victorian era provided.

Wilson

I greatly enjoyed Zack's analysis of the John Fowles novel. I admire his ability to dissect a text and find such relationships as Fowles's probable intention to use his nineteenth-century characters as cultural fossils. In my scientific writing I generally am striving to avoid ambiguity and attempt to assure a single interpretation in my papers, impossible as that may be. In literature, often the opposite is intended. I envy Zack's ability to read and weave various meanings into the words of others.

As Zack suggests, I do not believe that scientific Darwinism can appropriately be used, as some have tried, to shore up claims of social Darwinism, or to justify particular extremes in capitalistic economies. These are but examples of how individuals will try to extend a scientific theory beyond its limits, with no good evidence or justification.

Even Fowles appears to be guilty of misusing the idea of Darwinian fitness in places, such as when he claims that Charles "belonged undoubtedly to the fittest" (134), meaning, in this case, merely that he was a member of the upper class, not that he was fit in a Darwinian sense. Or again as Dr. Grogan asks: "'Have you read Malthus?' . . . 'For him the tragedy of *Homo sapiens* is that the least fit to survive breed the most'" (180). This, of course, turns Darwinian fitness on its head. Finally, in combining the idea of survival of the fittest with the idea of a need for adaptation and change in the individual, Fowles is stretching things. In Darwin's natural selection, it is not the individual who changes and evolves, but the population.

There is one place in *The French Lieutenant's Woman* where Fowles makes a point that will become quite relevant in our next chapter. Fowles interjects the following:

> Darwinism . . . let open the floodgates to something far more serious than the undermining of the Biblical account of the origins of man; its deepest implication lay in the direction of determinism and behaviorism, that is, towards philosophies that reduce morality to a hypocrisy and duty to a straw hat in a hurricane. (99)

While I do not totally agree with this statement, identifying humans as evolved entities, with body and mind located totally within the physical world, brings with it some profound implications, as we will see.

As a final note, another of Fowles's novels, *The Collector*, left me much more uncomfortable. To the extent that the chief protagonist in that novel can be likened to a scientist, the desire for control, in this case over an imprisoned woman, presents a disturbing picture. While many scientists are driven by a desire to understand the world and ourselves, some are interested in control, be it of nature or of other humans, and Fowles certainly paints the dark side of that desire.

6

Uncertainty

Part One: Uncertainty, Bias, Determinism, and Free Will

Wilson

There are many different kinds of uncertainty in science that will be explored below. The exploration will get us into some interesting issues related to whether events in the universe are predetermined or subject to control by human decisions and actions. The tentative answers science offers are disturbing and challenge basic ideas about what we humans are all about.

Changing theories

There are, and always will be, uncertainties related to scientific knowledge. Even Newton's laws, for three hundred years held to be truly accurate, proved subject to change, and were changed, by Einstein. While theories are sometimes refuted, and even so-called laws are modified, there also appears to be a real continuity in scientific development. There is a growing body of data that scientific theories explain, and the growth continues even as scientific revolutions occur. The strength of the scientific enterprise also can be seen, for example, in the fact that engineers still use Newton's laws for building bridges, skyscrapers, and airplanes. The laws turn out to be a close-enough approximation to serve us quite well in many situations. What was more revolutionary in Einstein's overthrow of Newton's laws was the change in perspective, the change in ideas about space and time that Einstein identified. I will save much of this for later. A key point is that the uncertainty and incompleteness in our knowledge, because of the possibility of modification by new tests and experiments, is a real and enduring part of the scientific enterprise.

In fact this uncertainty is one of the things that separates scientific method and knowledge from other human knowledge systems. Compare, if you will, religion and science. Most established religions are built on a foundation that is considered firm and unchanging: each is based on faith. Faith in the existence of a god or gods who is/are concerned with the affairs and doings of humans is an unchanging part of the belief system of most religions. Most also accept unquestionably some form of life after death—heaven, hell, or reincarnation, as examples. While, with time, a particular religion may change the details of its system of beliefs, a core of belief (in the existence of a god, for instance) cannot change.

Science isn't the same. There is no firm, unchallengeable foundation or set of assumptions or axioms. As we discussed earlier, the assumptions for one experiment can become the hypothesis for another. Consider such ideas as the description of space in Euclidean geometry; the idea of absolute time; the ether; the earth as the center of the universe; all organisms created at the same time, and unchanging; the universe as everywhere the same—uniform and unchanging. All these ideas served at one time or another as part of the foundation of science, largely accepted from the culture of the day—the basic assumptions that all humans had made at the time—and all were rejected when contradicted by experiments and data.

Even such fundamental assumptions as knowledge being attainable through observation, observed phenomena being subject to classification, and the idea of progress toward truth in science, are subject to testing and possible refutation. Obviously, if any of these fundamental assumptions, which underpin all of science, are found wanting, then science itself is in deep trouble.

Given that science is progressing toward truth, how close are we to it? We can't say. Is today's model getting us near a final truth? I think that we know some things beyond reasonable doubt, but we can never be fully certain. It is in the nature of science that the journey is never complete.

Could there be a radical reworking of our current views? Yes, that is possible, but in some subdisciplines in science it is becoming less and less likely, as I pointed out with the example of the periodic table. We will revisit this issue later, as we explore the nature of revolutions in science. But science must always live with an uncertainty about the final accuracy of our theories. This kind of uncertainty is part of what science is all about, and part of what sets it off from other kinds of knowledge systems.

Bias

One source of uncertainty in science is the possibility of bias in scientific experiments and analysis. There can be bias in individuals, in groups of individuals, and perhaps even in humans generally. All of us have biases because of the particular cultures we find ourselves in. This is easier to see when looking at the bias in other cultures, or in our own culture at different times, as is seen so obviously in *The French Lieutenant's Woman*. We also saw that there has been a history of such bias in attempts to measure intelligence in humans. Such biases are less likely to have an impact on physics than on such sciences as sociology because of the immediate cultural implications of sociological work, but that does not mean that physics is totally free of bias. There may be less of it in physics, and it probably is more subtle.

If we look back we can see that bias has affected the interpretation of scientific evidence and even at times impacted data collection. But we also can see how such bias may be corrected by later generations of scientists, as Gould did in his book *The Mismeasure of Man*.

It is not true that progress in science just consists of one set of biases replacing another; we actually get better at doing experiments as well. For instance, the measurement of skull volumes can have bias removed from it by doing blind experiments, where the person doing the measurement does not know, at the time of the measurement, the identity of the skulls being measured. Today in biomedical studies, one often does double-blind experiments. A good example of this is a study of some new drug, where neither the experimenter nor the subject knows who is getting the medicine and who a sugar pill. Such information is kept from both until the end of the experiment. Such bias-reducing approaches as double-blind studies are routinely followed today when there is a risk of unintentional bias. Individual scientists are as subject to bias as anyone else, but the enterprise of science values objectivity, and works toward it by employing corrective mechanisms at a given time, and across time, as cultural bias becomes more evident. In addition, nothing exposes bias more effectively than a well-designed experiment that demonstrates something counter to the biased view.

Is it harder to do experiments without bias in, say, sociology than it is in an area like physics or chemistry? Yes, undoubtedly. There are more uncontrolled variables, and sometimes even uncontrollable variables. These give more opportunity for bias to creep into the interpretation of experimental results. There is a greater potential for cultural bias in some experiments, such as determining whether social welfare is a good thing, than in others, such as testing whether a particular nerve cell makes a particular protein. Bias can influence experimental design, data collection, analysis, and interpretation in ways that are sometimes subtle—and sometimes not so subtle. Even in simpler studies of the opinions of individuals, one has to be very careful about the wording used in questionnaires, because changes in wording can make a big change in the outcome. Compare "Are you in favor of killing unborn babies?" with "Are you in favor of women having a choice of whether to terminate a pregnancy?"

Would Martian physics be different from ours? Certainly it is conceivable that there are good models of the world other than that of our present-day natural sciences. But were we to compare notes with a scientist from Mars, one or the other set of paradigms would likely prove to be superior in accounting for all of the evidence and data, or a mix-and-match among the theories might do a better job.

But there have been some rather outrageous claims made recently by some in "feminist science" that assert the necessity of a rewriting of much of science.[1] To the extent that someone could develop a better set of theories to account for the data, more power to her or him. But until such a new system is developed and demonstrated to be superior to what we now have, it seems silly and arrogant just to make such an assertion. Some have argued that even the terminology of physics: force, power, and so on is somehow too masculine. But were the words to change, the general laws wouldn't. F = ma. One could change the wording, but the way that bodies move about in the universe isn't going to change, and so a law describing such motions will have to be similar to what we now have. Finding a better law is no easy task, and until such is found, the argument that radical changes in physical laws are needed rings hollow.

There certainly are ways in which science, until recently, has suffered from deficiencies due to its being done more by men than women. For instance, some of the studies of human life styles and health, which had only males as their subjects, were not as useful as they might have been for determining whether particular life styles would benefit women as well. This deficiency has now been recognized and the change has resulted in many more studies that include both males and females. But that kind of bias in subject selection did not invalidate the science—it just meant that the results of certain studies pertained to males, and that there was insufficient evidence to say whether females were affected in the same way. The story was incomplete, but not wrong.

Perhaps more tellingly, feminist perspectives have been a help in studying, for example, other social animals. In field studies of primate behavior and primate social systems, male scientists sometimes appear to have had blinders on concerning their data collection and interpretations. For instance, assumptions were made about the importance of male aggression in holding groups together, with the role of females being underplayed. It also was assumed that dominant males within a primate group fathered most of the offspring. Alternative possibilities were tested only after female scientists proposed other models, and indeed it came out that the female primates in the studies did mate with other males from outside the group and were important in group dynamics. Here, the science of primate behavior was improved by the new perspectives and more open viewpoints. More generally, different perspectives can be useful and enlightening and should always be welcome in science. Thus far, these changes have been important to some areas of science but have not been radical for all of science.

Had human beings and human culture evolved and developed differently, and had women historically been in the majority among scientists, I don't think our scientific world view would be all that different today. Science would just have started with a slightly different set of biases, and then the addition of more males among scientists would have necessitated a similar number of modifications, adjustments, and improvements like those we are seeing today, as more women enter various areas of research.

Perhaps a more significant general issue would be whether science is biased because it is being done by humans. In part this question involves answering another question: Are there general biases shared by all humans in all societies over all time that act as blinders, keeping us from seeing certain things or recognizing an error or limitation? That is tough to answer. There certainly were biases in our beginning assumptions (such as Euclidian geometry) that have *not* limited us, and science has gone beyond such limitations, as we have already discussed.

Perhaps if there are aliens who have produced an alternative scientific system, we will be able to compare notes with them to seek an answer, for us and for them. But even then, one could ask about biases or limitations we share. In the end I think the answer will be found in how closely we can come to understanding the universe and our place in it.

Granting the limitations that exist because of the kinds of uncertainty described above, we can ask how well science can model or predict actions in the universe, including human actions.

Determinism and Free Will

How fixed, or determined, is the universe? Is the future course of actions in the universe predetermined? If so, can science hope to be able to predict that future course? Also, if the future is either predetermined or unalterable, what are we to make of our sense of free will? By free will I mean the freedom to make decisions and take action in the world independent of antecedent physical causes of any kind. Does such free will exist? To work toward answers to these questions, we will need to explore several possible sources of indeterminism. We also will need to examine, to some degree, the role of consciousness in human decision making.

I will begin with a classical physics perspective and then bring in new issues raised by modern physics. Classical physicists of the 1700s and 1800s appeared to accept an absolute kind of determinism, at least in principle: if one knew all of the initial positions and directions of movements of all of the particles in the universe, and knew the nature of all the forces operating in the universe, then the future course of the universe could be determined.[2] In this classical world of physics, all matter was viewed as billiard-ball-like. One could describe the initial conditions and use Newton's laws of motion to predict the future of the entire system. Newtonian determinacy was a model that pointed to a clocklike universe. The future motions and collisions of every particle were predetermined by its initial position, momentum, and the forces that acted on it.

However, it certainly was realized that for practical reasons, gaining enough knowledge to predict all aspects of the future was unattainable.

Even with today's list of forces acting in the universe, which we think is fairly complete, there just are too many objects ever to have a full description of an initial state for all the atoms and molecules. Modern physics has given us other reasons to question the simple view of a fully determined universe. So let's explore some of these various kinds of limits to predictability and determinism. As we do I will try to keep in mind the questions concerning free will, and look for any way that the limits might give us an opportunity to exercise free will in our actions.

Uncertainty Due to Complexity

We know that we cannot determine the position and momentum of all of the particles, even in a small volume of gas, simply because there are so many of them. It would take many lifetimes just to try to develop such a catalog, and the particles would be moving as we attempted to complete the list, so we'd never finish the task.

But this kind of inability to determine the starting conditions is just a complexity limitation, not a limitation in principle. We know that atoms and molecules are colliding in a gas, and while we can talk about the random nature of those collisions, that randomness is at another level predictable, in principle, given Newtonian mechanics. We have learned to deal with this kind of lack of predictability by considering the molar properties of the gas: temperature, pressure, volume, and so on. This kind of solution to complexity indeterminism works for many of our purposes. For instance, to determine what an automobile is going to do, we don't need to know the position of every molecule in the vapor in each cylinder within the engine. To predict the behavior of the automobile, it is sufficient to know generally about the fuel burning, to know that the spark is occurring at a particular time in the cycle, the direction the wheels are pointing, and so forth.

Now let's consider the human brain. Here we have ten to a hundred billion nerve cells, or more, most with thousands of inputs and outputs. We cannot ever hope to work out an initial state for the neurons in a single human brain, any more than we could the molecules in a volume of gas. Furthermore, each of us has some unique wiring connections and strengths among our neurons. Thus, we will never be in a position to predict fully the future behavior of a human, even under classical physics. Neuroscience indicates that our decisions may be made of, and at least depend upon, the electrical activity in our neurons. Whether a particular neuron fires an electrical impulse at any instant depends upon the instantaneous input to that neuron from, typically, tens or hundreds of other neurons. Such firing of electrical impulses can occur at rates up to five hundred times a second in any single neuron. The complexity is extreme.

However, as with the molar properties of a gas, all of the complexity in human brains and the limitations on predictability that it imposes have not stopped us from learning some general principles about how humans behave, and how the brain produces such behavior.

It has been pointed out by others that determinism need not entail predictability.[3] The world might be fully determined and yet we might never be able to predict it. Our limited ability to make predictions does not mean that the world is not determined. As molecules combine to make macromolecules, and macromolecules combine to form organelles, and organelles work together as parts of neurons, and neurons team up to become brains, there is no evidence that the resulting structure or its complexity frees the behavior of the more complex systems from that predicted (or inferred) from its components. The laws of physics and chemistry appear to give a complete description of the forces that govern the actions of the atoms and molecules of human brains. There is no evidence so far of new forces or emergent, vitalistic phenomena in brains.

Chaos

There is a special kind of complexity that I'll consider briefly.[4] Systems exist where very small changes in the starting conditions can bring about big changes in a later state of the system. This is called *chaos*. Chaos underlies our inability to predict what the weather will be like one year from today, except in the most general of terms. A classic example of chaos, whether accurate or not, is that of a butterfly in West Africa flapping its wings and initiating wind flow that eventually results in a hurricane in Florida—which, without that butterfly, would not have occurred. Small disturbances, such as air movements caused by a butterfly moving from one flower to another, obviously do not disturb the air very much. However, such a small perturbation might shift wind currents just enough to make the difference, ultimately and after many steps, in whether a storm develops off the coast of Africa. That storm could then

become a hurricane. If the butterfly stays put on one flower, in the absence of the little air disturbance, no hurricane develops. In spite of its
implausibility, this example may help you get the picture. Some systems
are chaotic: changes in the starting conditions too small to be measured
accurately lead to totally different outcomes.

Chaos really is a special kind of complexity limitation on predictability. It is not a question of violation or inadequacies of any physical laws,
just of small differences in starting conditions leading to big differences
in outcomes. It is interesting to consider whether our brains might sometimes work this way. They well might, at least at times.

Both for chaos and for complex systems with many interacting units,
one really is confronting a limitation on our ability to determine starting
conditions or to follow complex actions. Because of complexity and chaos,
we probably never will be in a position to predict the future actions of a
human in complete detail, but this alone does not prevent human brains
and mental activity from being fully determined. Complexity alone does
not suggest that violations of physical laws could slip in, or that the unfolding of human action isn't fully determined by physical laws—the
same physical laws that appear to govern all activity within the universe.
There is, however, another form of uncertainty that may limit complete
determinism.

Uncertainty Due to Quantum Mechanics

Modern physics has introduced some revolutionary new views about the
nature of matter.[5] Some of these views appear to impact determinism in a
more direct way. They seem to limit our ability to determine the future,
to predict outcomes. What upset the apple cart during the last century
was quantum mechanics. Before quantum mechanics, matter was viewed
as quite solid, and fixed in size, rather as a billiard ball or rock appears to
us. We now believe that matter, far from being billiard-ball-like, pos-
sesses a wave nature, something like the wave nature of light, and so is
not as solid or fixed in size as it appears to be in our perceptions. It is not
that our perceptions are so far off for objects the size of billiard balls and
rocks, but that on a fundamental level, even the matter in such objects, if
examined very closely, has a wavelike character as much as a particle-like
character. Furthermore, light, which had been thought to be purely a
wave phenomenon, has a particle nature as well as a wave nature. So,

there is a wave/particle duality that holds for light and for all elementary particles, such as electrons and protons, and even, in a limited way, for larger objects.

There are two consequences of this that I want to discuss. Quantum mechanics states that it is not possible to define both the exact position and the exact momentum of an electron or atom at a given moment. To the extent that we know one with greater certainty, the other becomes less precisely defined—position or momentum. For our purposes, momentum can be taken simply as mass times velocity, so we can say that knowing the precise position of a particle at the same time as knowing the precise speed and direction of movement for the same particle is what is limited. This can be viewed as resulting from the act of measurement: as we attempt to define, say, position, very accurately, we disturb the momentum of the particle. However, some see quantum mechanics as more fundamentally limiting the "real" position and momentum of particles, independent of the measurement effects.

Whatever the proper interpretation, the uncertainty is very small for objects the size of tables, chairs, humans, ants, and even bacteria. For these we can safely, almost always, ignore the uncertainty. But this uncertainty cannot help influencing us as we think about determinacy. There is an inherent uncertainty in physical events, and even the smallest uncertainty at any moment can grow to yield a larger uncertainty with time. If we are a little uncertain about the position or direction of a single molecule in a gas, then after time and collisions with other gas molecules, whose positions or momentums also were slightly uncertain, we will not be able to predict where the particular molecule will be found in a container. That will not usually affect the molar qualities of the gas significantly—the gas will still have a particular temperature and pressure, as the molar properties of the gas result from statistical averages of very large numbers of molecules in the container. Many accept that quantum mechanical uncertainty eliminates any fully deterministic view of the world. No longer can we predict, even in principle, the exact course of future events in the world.

There is a second issue that arises from the wave/particle duality of matter. We should now more accurately describe events, or options for events, at the quantum mechanical level, as occurring with certain probabilities. This does not have much impact on objects the size of humans—

we don't see ourselves as blurry waves unless we've been drinking too much. It may not even concern single cells significantly, at least for most of their actions over shorter periods of time. However, if we look at an electron in a single atom, the uncertainty is very evident.

The uncertainty can be seen, for instance, if we fire electrons toward a barrier that has two narrow slits, close to each other, in it. For the few electrons that pass through the slits, we no longer can tell which slit the electron goes through because it behaves as if it is influenced by both slits! If we place a screen on the other side of the two-slit barrier, we get a diffraction pattern of electrons hitting the screen, rather like what we would get if we shined light at the slits. The electrons are hitting at a variety of points behind the screen, with no obvious peaks right behind the slits. The maximum number of electrons are found at a point on the screen behind the point that is right between the two slits. The electrons are exhibiting properties that are wavelike. In a sense, each electron goes through both slits, even though we formerly thought of electrons as little, solid bits of matter, they are acting as if they are smeared over a larger range of space because of their wave nature. We cannot predict exactly where an individual electron will end up on the screen, but can predict the probability of its landing at any given point.

So there is a crack in the deterministic view of the universe. Where does this uncertainty and delocalization get us? It is crucially important to realize that quantum mechanical uncertainty is random in an absolute sense of that term. It is a basic principle of quantum mechanics that actions such as electrons passing through the two slits, or the decay of a radioactive atom, cannot be predicted or influenced. In the case of a group of similar radioactive atoms, the average number of decays over time in the group can be predicted, but the time of decay for any given atom is random and probabilistic. Similarly, the wave nature of an electron contributes to uncertainties with respect to where it might be located at any instant, but the probabilities are set and cannot be adjusted except by other physical forces or objects. Any other adjustment would violate quantum mechanics.

Thus, quantum mechanics appears to impact our ability to determine the future, to predict outcomes. There is a lack of full predictability, as future events at the quantum mechanical level become probabilities rather than certainties. And even though the uncertainties are small, over sufficiently long periods of time the uncertainty grows. We certainly can acknowledge a lack of full predictability, and some would see quantum mechanics as presenting us with a lack of complete determination.

However, for free will, it is not very helpful to have this kind of a lack of determinism. What replaces determinism here is merely chance. Furthermore, within the time frames of human decision making, and given what would be required to trigger the brain activity underlying human actions and speech, quantum mechanical uncertainty does not give us much to work with.[6]

For free-will proponents it would be more satisfying for humans to be able to influence outcomes beyond physical law and chance. Is there not some way for each of us to govern what happens, to govern physical events so that our actions are pointed in the direction that we wish? As yet there are no data suggesting such an ability. Furthermore, if we had this kind of control—if this were how free will controls brain action—it would violate physical law, because the quantum mechanical probabilities would be altered. Quantum mechanics would no longer be true.

So, for the purposes of free will, it does not appear to be very helpful just to be free of determinism if what replaces it is merely chance. Indeed, the fact that our will is so seldom thwarted suggests that our brain's decision-making machinery is not much affected by quantum-mechanical uncertainty or probabilistic events. To the extent that we make decisions, we appear to rely on deterministic events in our brains to carry out that will, as our brains generate actions through control of muscle contractions.

Chance and Determinism

It might be worthwhile to clarify a couple of the ways we use the word "chance." I have been speaking of chance, probabilistic events under quantum mechanics, but there is another way, unrelated to quantum mechanics, that we often speak of chance events. Many of these more general, chance events are happenings in the universe that are predictable, at least in principle, by physical laws. For instance, consider the chance event of a meteor that we now believe hit in the Yucatan about sixty-five million years ago. That meteor is thought to have resulted in the extinction of dinosaurs. That chance event was certainly predictable by physical laws—we are pretty good at determining the paths of larger objects in the solar system, even many years ahead of time. This particular chance event appears to have changed the course of evolution on earth, perhaps even leading to the evolution of thinking beings, as mammals were better able to survive in the absence of dinosaurs.

So, what we call chance events can be of different sorts with respect to whether they are predictable by physical laws. And what we call chance events of both sorts can have a big impact on humans. As an example of quantum-mechanical chance events with the potential for a big impact, consider the variations, or mutations, which also contribute to evolution and its direction. At least some of these mutations are produced by chance events of a quantum-mechanical sort. Consider cosmic rays, which pass through our bodies every day. They are high-energy electro-magnetic particles. Occasionally cosmic rays will interact with atoms in our bodies and make changes in molecules, including DNA, which can produce a gene mutation. Because such interactions are at quantum-mechanical levels, we could only predict on a probabilistic basis whether a particular gamma ray would interact with a particular DNA molecule in a particular cell, even given the most complete possible knowledge of the gamma ray and the DNA. But the resulting mutations are the variations on which natural selection acts during evolution. Humans appear to have evolved on earth because of many chance events of both the meteor sort and the cosmic-ray sort.

Furthermore, consider each of us and our own uniqueness. Consider the odds involved in our particular egg developing in Mom in that particular month; the odds of our one sperm developing in Dad; of the one result from the "night-sea journey" Each sperm differs from the others, no two are likely to be exactly alike in their genetic makeup, and the same is true for the eggs. What was the chance that you or I came to be, and not someone as different from us as a brother or sister? And what of the complex decision of our parents to have intercourse on that particular night, at the particular moment when that one egg was ripe and that one sperm was there and positioned just so? The chance events involved in the night sea journey are many, and some may be chaotic. Any attempt to describe fully the event from a physical standpoint would fail from the complexity: the thoughts of the pair of humans leading to intercourse at a particular moment, the equations for flow of sperm, and all the rest.

As perhaps we can appreciate now, some of these events are determined within the limits of physical law but are beyond our ever working out because of complexity. Others are chance events in the quantum-mechanical sense, not fully determined. However, in all of this, do notice that, to the extent that determinism is not total, it appears to have been supplemented only with chance.

Conscious Decisions and Free Will

So we are left with a puzzle. If everything that occurs in the universe is either determined or probabilistic and due to the result of truly chance events, then where does our feeling of having free will fit in? Most of us consider ourselves to be capable of willing, of performing volitional acts in the world. We feel as if we make a difference, as if we are autonomous beings using our own free will to decide what to do.

Some have interpreted modern neuroscience as sending us another possible message about whether we actually make conscious volitional decisions or take conscious action. Much of brain activity and brain processing is not at a conscious level. There is considerable brain processing that now is known to occur before we have experiences. For a simple example, consider visual-information processing. There is considerable processing of information in our visual cortex, at the back of the brain, of which we are not directly aware. We only "see" the image of the world as if we were looking out from our eyes. We are ignorant about much of the processing that gives rise to the visual perception.

Some neuroscientists have taken the fact that much brain activity does not occur at a conscious level as indicating that our consciousness is an epiphenomenon, an aftereffect that is added on to decisions already made by preconscious brain activity.[7] In his book *Consilience,* E. O. Wilson appears to agree, as he suggests that our sense of free will is a result of our ignorance about such preconscious or unconscious processing in the brain, and that we are left with an illusion of conscious control. But I have previously argued that epiphenomenalism is not a plausible view of the relationship between mind and brain.[8] Owen Flanagan also has offered an extended argument against it in his *Consciousness Reconsidered* and more recently Carins-Smith has summarized evolutionarily based counter-arguments in his *Evolving the Mind.*

I suspect that the conscious mind plays a major role in decision making just as an executive in a large organization does: the executive is not directly aware of much of the detailed activity within the organization, and there often is prior activity by others before the executive obtains the relevant information and makes a decision. The executive also does not make all of the decisions made by the organization. Nevertheless, the executive does make decisions, and often the more important ones for the organization.

Neuroscience has not established an illusional, epiphenomenal role for mind and consciousness, and the evidence that much brain activity is not conscious in no way diminishes the likelihood that conscious decisions can cause actions.

Where does this all leave us with respect to human actions? Science seems to point in this direction: we can and do consider evidence, situations, and alternatives before we act. It may be predetermined, within quantum-mechanical uncertainty and randomness, what will happen and what decisions each of us will make, but that does not prevent us from making the decisions. Our brains are, in a sense, decision-making machines. As we make plans for action, we consider a variety of factors, moral and ethical among them. Emotional as well as logical influences play upon our decisions. Our frontal lobes appear to be special brain regions for planning, and for mixing and blending the various factors that must be weighed—the options for action, the possible reactions of others to particular acts we might make, and the consequences for ourselves and others.[9]

Our legal system presupposes that we are free to make choices. If everything is either predetermined or a result of chance, then what do we make of our deeply held beliefs and human cultural systems? As was said in *The French Lieutenant's Woman,* do we "reduce morality to a hypocrisy and duty to a straw hut in a hurricane"? (99). We need to confront the likelihood that even our seemingly freely made decisions have outcomes that are predestined, within chance limitations. This view rests on modern science and certain scientific views of the relationship between mind and brain, which we will explore in later chapters.

What might this mean for our legal system? Even if individuals do not have free will, they still make decisions. Perhaps we should not think of prisons as existing to inflict punishment, but instead should stress the dual role of attempting to influence inmates' future actions and of preventing them from repeating the actions that got them there. That approach might bring about some positive changes, especially in the United States, where we imprison such a significant number of individuals.

Even if all human behaviors are predestined within chance limitations, neither we nor any other human has any way of knowing that outcome beforehand. Consider the complexity of a human brain: the brain is perhaps the most complex piece of matter in the universe. Should the possibility that all of what happens is predestined, subject only to the whims of random quantum events, influence how we behave or what decisions we make?

Science suggests that we may be determined, except for quantum-mechanical uncertainty and random events, by our genetics, environment, and experiences. As we develop, the individuals that we become and the decisions we make could be predestined. We are constrained by our genes and by our environments. But if one thinks about this a little, how could it be otherwise? We are who we are.

Is this a final view from science? It may be too early to tell. We still need to know more about mind and how the brain works to generate consciousness and exercise will. Until we have at least some idea about these issues, we must recognize some scientific uncertainty about the view that I have described. This is not one of the conclusions about which we can claim confidence beyond reasonable doubt. While we do not have a finished scientific analysis of the possibility of free will, the direction in which science seems to be leading us should be evident.

Part Two: Language and Uncertainty

Bowen

In my previous discussions, I have been developing the differences between scientific truth and the insights of the metaphors provided by literature. In doing so I have been borrowing more or less unconsciously from the work of Friedrich Nietzsche, for whom "truth" is composed of culturally conditioned metaphorization and hence is unstable and not a fixed reality:

> What, then, is truth? A mobile army of metaphors, metonyms, anthropomorphisms—in short, a sum of human relations, which have been enhanced, transposed, and embellished poetically and rhetorically, and which after long use seem firm, canonical, and obligatory to a people: truths are illusions about which one has forgotten that this is what they are; metaphors which are worn out and without sensuous power; coins which have lost their pictures and now matter only as metal, no longer as coins.[10]

As a teacher of language and literature, I am concerned less with absolute truth than with the verbalizations and metaphors that human beings have created to express it. In this prioritization I do not represent the humanities as a whole, least of all epistemologists like my colleague, Susan Haack, who has never given up the pursuit of truth, and feels that science is an extraordinary embodiment of rational inquiry.[11] So do I. But while I am not as despairing as Nietzsche of coming to grips with any absolute truth, nor as sanguine as the scientists or Haack in eventually achieving it, I am fascinated by the creative process which produces the narratives and metaphors that attempt to express it. Primary in that process is language itself.

Language and its hopefully resulting *communication* are both slippery terms with which to come to grips. For purposes of this discussion I use the term *language* as an arbitrary structure of verbalization intended to convey meaning (*communication*) to others. There are forms of communication other than speech and writing that would not fall within the narrow definition of language I propose. Alternative communication patterns include pictures and other nonverbal material wholly outside but often used in conjunction with verbal speech acts. For instance, the hissing, meowing, and purring of cats all fall into my definition of language because they are all speech acts, but feline body action—their tail waving, back arching, and rolling over on their backs—works just as effectively to express their feelings (aggressive, playful) to their contemporaries as well as their owners, who are trained to interpret their body language. But while human beings rely primarily on speaking and writing to communicate, this does not say that nonverbal communication does not exist in our species. Bodily or physical communication is just as common in people as it is in other animals, although people frequently use it unconsciously.

Kinesics, or bodily gestures of hands, eyebrows, leg movements or positioning, are only a part of the multifarious signals we constantly use to accompany and modify or emphasize our verbalizations. Territoriality often plays a large cultural role in human communication. Distance sensitivity is largely subsumed in cultural behavior, although we do not necessarily recognize it. For example, we may characterize someone as speaking with a foreign accent even though the sound of her disembodied voice admits no trace of it, simply because her gestures or stance bespeak another culture. In some cultures people stand almost nose to nose with each other, and when the person of another culture to whom they address themselves unconsciously backs off a bit, the speaker may think the listener "standoffish," and conversely the recipient thinks of his opposite as "pushy." Hugging, touching, and kissing in greeting casual acquaintances are approved in some cultural circumstances and inappropriate in others. The territorial specifics may either be unconsciously culturally acquired, or consciously studied and learned for cultures other than one's own, in the same way we master any new language.

The degree to which language itself carries heavy emotional connotations depends in some measure on whether we identify it as being simply an arbitrary signifier or the signified essence itself. Even in Genesis, God gave Adam the right to choose the names of the other creatures, thereby establishing some sort of arbitrary control over them. The naming process conveys at least a conviction of superiority for the namer. But naming creates only a fictional and highly arbitrary relationship between namer and named, and not a direct confrontation with reality. A lion, even if I called him a beetle, could still eat me if he chose. The point is that the significance of any linguistic construct lies in some combination of what it communicates for the originator and what it means for its ultimate recipient/audience.

Therein lies another whole can of ambiguous worms. At the far end of the communication spectrum is the listener/audience who receives and reinterprets the language (words) in the light of her own conceptual framework. Reception theory and reader response deal with the way in which the received message is translated by the recipient. The act of reception is if anything more complicated than the encoding process itself. How we perceive what we hear or read is conditioned by nearly every memory and association we have experienced or formed throughout life. What might be pejoratively called our prejudices might also in other circumstances be termed our education, our past experiences with the subject, or our general social/cultural/familial conditioning. In nearly every set of circumstances the recipient reader/audience has some conditioning for receiving all imaginable signifiers. For example, for years I have received letters from Publishers Clearing House promising that I am only separated from winning ten million dollars by mailing them back their forms. The Clearing House banks on the supposition that the some of the applicants who actually fill out the forms are conditioned to think that their chances are enhanced for winning a prize if they order magazines, even though the firm is legally enjoined to print (in microscopic type) somewhere on the pitch that those who order nothing are still eligible to win. I am conditioned, on the other hand, to throw away such letters and illegally to screen my wife's mail and throw her temptations in the trash, since I react adversely to stacks of expensive unread periodicals on our coffee table.

I listen to talking heads on television to hear pundits put right or left political spins on every bit of news, and am fascinated by how far from what I consider reality their opinions and interpretations are. The skewed representations of day-to-day news must be modified by a knowledge of who owns the particular reporting source and which advertisers or political organizations ultimately pay the reporters' salaries for various networks, think tanks, journals, and publishers. Similarly, a whole analogous, but perhaps less formal process dominates our perception of every utterance, from the pronouncements of someone we know to be an idiot, to someone with his or her own unique agenda. Such filters may become a source of what we will later discuss as "paranoia" in connection with *The Crying of Lot 49*.

Reception theory deals as much with general perception as it does with the cause-and-effect distortion discussed above. I make it a habit to re-read those student papers originally graded in the morning, when I had gotten up cranky, or marked when I was exhausted late at night, because there is for me a recognized subjective bias built into grading essay papers, one not encountered in the more prescriptive scheme of fill-in-the-blanks or true-or-false preconceptions graded by template. I don't reread from sheer altruism, but because I know that under other, better, circumstances, student opinions can sometimes be judged far more favorably.

The relation between what is signified and its arbitrary signifier is complicated by the fact that both the sender and the receiver of the message invest its language, the signifier, with all the realistic properties of the thing signified. We often make no difference between the word itself and what it represents, or signifies. That sort of confusion is endemic to all signifiers, verbal or otherwise. For example, congressmen may indulge themselves in long, heated, meaningless debates on the meaning of flag burning and whether to incarcerate people who do it or even propose to do it. The word *flag* is twice removed from what it signifies. "The flag" is an arbitrary signifier for an even more arbitrary metonymical connection between *flag* and our nation—a signifier that changes every time we add another state. Words themselves are equally subject to rapid change due to politics or other circumstances of what they represent. Political/social issues such as abortion become enmeshed in a snarl of connotations like murder and baby killing versus women's rights to their own bodies. Or pleasant-sounding signifiers like "pro-life" and "pro-choice" are favored as the arbitrariness of the signification process becomes the battleground for recruitment of adherents to opposing viewpoints. No one buys used cars any more, they buy "pre-owned" ones.

The relation of words to meaning has long been the subject of a number of theoretical schools of literary/philosophical study. I will mention only three major ones: semiotics, structuralism, and deconstruction.[12] One of the originators of what became semiotics was Charles Sanders Peirce, whose work began in the 1860s, although it did not appear in book form until 1923. We shall hear more about Peirce later in the *Crying of Lot 49* discussion. Peirce believed that reality is "what is real for us." He goes beyond empirical philosophers by insisting that thought is composed of signs, in effect that "Cognition equals Expression." Peirce does not hold to the Neoplatonist view (to be discussed in a later chapter) that things do not exist prior to being thought of by someone. Rather for him their significance relies on their becoming a part of a sign process. According to Peirce,

> A sign stands *for* something *to* the idea which it produces or modifies. Or, it is a vehicle conveying into the mind something from without. That for which it stands is called its *object:* that which it conveys, its *meaning;* and the idea to which it gives rise, its *interpretant.*[13]

The *interpretant* is the idea formed in the mind of the receiver of the message, its interpreter. The field of semiotics concerns itself with "the collective total of all that is in any way or in any sense present to the mind, quite regardless of whether it corresponds to any real thing or not"(*Papers* 1:284)

The idea was refined by Ferdinand de Saussure, who introduced the system of cognition that still has currency, even in the present discussion: "Language could only be understood as a system of signs, the study of which he [Saussure] called semiology. The sign consists of signifier (the sound image) and the signified (the concept, the thing). But it is a property of language that 'the bond between the signifier and the signified is arbitrary' (*Course in General Linguistics,* 67)."[14] The entire lexicon and grammar system as internalized and understood by native speakers, Saussure calls *langue.* While Saussure defined part of the process of communication he does not give any space to the interpreter's role in the recreation of meaning.

   That task was taken on by the structuralists, who began historically by seeking common threads in narratives and inferentially and then statistically creating formulas in which they occurred in oral and verbal storytelling, tracing them through history to attempts at human understanding of the environment,[15] and eventually postulating a system of signs which includes such nonverbals as clothing, kinds of food, and so on, embracing all of human tribal activity. It is in these theories that cultural anthropologists and contemporary literary scholars find common ground. Remember how Fowles used such social conventions in *The French Lieutenant's Woman*?

To a certain extent Jacques Derrida undermined the regularization of overarching identifiable "laws" governing the semiotic and structuralist approaches to systematic understanding in human communication through the ages and in current cultures. He sought to destabilize the illusion of truth, stability, and knowledge on the ground that they are formed on words that themselves are polysemic—hybrids that are both one thing and another, as well as the spaces between words (their absence) which is a part of the whole construct of thought. This *différance*, to use Derrida's term, is all-pervasive in that its metaphysical qualities infuse every utterance, the unspoken on which we base a cognition, which is always partial, and ultimately indefinable, precisely because it contains the presence of absence. It makes puns based on multilingual combinations possible, and infuses them with possibilities undreamed of, as well as those intended by their communicators. The act of reaching outside the system to inform any interpretation of the text is a nonliteral one, and it has led many to fear the destabilization of meaning for any communication, even as we realize its multifarious potentiality for a plenitude of meanings.

So arbitrary are the signifiers of language that scholars like Derrida seem ready to abandon any absolute belief in the results of the signification process itself. But even traditional historical linguistic analysis indicates more mundane sources for uncertainty in human communication. I will use the English language as my principal example, hoping that by some Derridian metaphysical process you will apply it to others.

Living (currently spoken) languages are characterized by a constant process of evolution in which a number of new words and signifiers are added daily, as well as equally dynamic changes in existing words/ signifiers. Each discipline or social group employs a distinctive vocabulary to distinguish issues important to its adherents. This is not confined to medical, scientific, or humanities disciplines, but to all manner of cultures. For example, in societies less interested in hereditary kinship factors, most people don't know the precise meaning of what being a third cousin twice removed represents, while other cultures in which familial relationship is of paramount importance have a special word that everyone understands for that particular relationship.

Literary writers whose interest and creativity are concerned with philological matters often capitalize on antique definitions of words still current, but with evolving connotations, because often the origins and former definitions of certain terms modify and enlarge present connotations and thus afford reinvigorated meanings to words in present common use. That is why for my discipline the *Oxford English Dictionary*, with all its historical etiology and historically serialized examples of use in literature, is of such importance to literary and linguistic study.

Language systems modify individual words/signifiers by attaching bundlings of sound, words, or spoken inflections. Words are divided into phonemes or single-vowel bundles, preceded by other signifying bundles to reverse or modify the principal root. Prefixes like "un" or "ir" or "a" usually reverse the meaning of the root word, while suffixes like "ing" can change verbs into nouns, and other suffixes can change singular to plural or change tense from present to past, and so on. The form that all of these take is peculiar to a given language or language system, and is just as arbitrary within that system as the naming process itself. English spelling, for instance, was formalized only recently—in the eighteenth century—and its grammar, evolved and loosely constructed from Latin forms applied to a basically Teutonic language, was largely codified not by linguistic experts but by early printers, who felt that their multiply produced pages of print demanded standardization. Latin-derived languages commonly used suffixes to denote the relation of substantive forms (names/nouns) to the meaning for which they were intended—the man, of the man, and so on, while Teutonic languages mainly used the word's position in the sentence to denote whether it was the active subject or object of the action. For example we understand the meaning of the sentence, "Jack hits John," because of the positioning of those names rather than any modification in their spelling. Difficulties arise when a Latin grammar is applied to nouns, verbs, and other parts of speech, which often must treat action words (normally referred to as verbs or verbals) as nouns, creating a hybrid confusion in such examples as "To run is to fall" which requires normal verbals to be translated respectively into the subject and object *acts* of running and falling—the second state/noun caused by the first. The point is that the more precisely catechistic the English grammar teacher tries to be, the more incongruities with which she has to cope.

I obviously haven't given up on language as meaningless, since we all have to use it regardless of its overabundance of imprecision and variation. So far I have only talked of the difficulties in our own language, but these are compounded by the enormous number of foreign languages currently spoken. The composers of Genesis tried in the tower of Babel narrative to rationalize the perpetual problems people have with the evolution of disparate, mutually unintelligible tongues, while linguists have continually attempted to uncover the discrete mechanics of the multitude of languages throughout the world in order to begin the process of translating the more important of their documents into languages familiar to us. Keys to decode the systematic bases of seemingly lost languages are still being uncovered, like the Rosetta Stone, discovered only in 1799, with its inscription in three forms of writing, which provided the basis for deciphering the ancient Egyptian hieroglyphs.

Spoken language and the literary language based on it are in a constant state of evolution in both time and space. Variations grow out of the spatial separation of individuals, from dialects to foreign languages, resulting in every living language being in constant flux. Translations, no matter how assiduously they are executed, never provide all the nuances of the original in such a way as to duplicate exactly the information conveyed by the original. And the problems inherent in a translation's capturing the sense and spirit, as well as the literal equivalent, remain the objects of international translation symposia today.

These tedious examples are given merely to indicate the imprecision of language in relation to the external phenomena they try to convey. We all, in our conscious minds, predominantly think and reason in language. We internalize its arbitrary definitions, grammar, and structure in our own rationalizing process, and in our conscious opinions and behavior. Arbitrary as it is and mistaken as it may be, language plays a major role in formulating our consciousness of the world we live in, and it does so in the most fundamental ways. Consider the opinion of first-grader Stephen Dedalus, the protagonist of James Joyce's *A Portrait of the Artist as a Young Man:*

God was God's name just as his name was Stephen. *Dieu* was the French for God and that was God's name too; and when anyone prayed to God and said *Dieu* then God knew at once that it was a French person that was praying. But though there were different names for God in all the different languages in the world and God understood what all the people who prayed said in their different languages still God remained always the same God and God's real name was God. (16)

While the above may represent merely the musings of a precocious child, no one can deny that when Stephen became a man he put away a lot of childish things. Stephen/Joyce was to become the linguist who wrote *Finnegans Wake* in seventy different languages, and even internalized French, Italian, and German to the point of thinking in those tongues, but few us ever reach the mature Joyce's stage of linguistic fluency. However irrational our monolingual perception may be, it still governs our rational logic, thought patterns, and ultimate meaning. The danger therein is that, like young Stephen, we may not perceive these patterns to be arbitrary or subject to interpretation.

The dismissal of President Clinton's deposition during the Monica Lewinski business stemmed in some measure from what seemed like hairsplitting in his qualification of the verbal testimony to the effect that it depended "on what *is* is." The House committee and the TV pundits made hay of their own linguistic ignorance by not seeing that there was a point to defining precisely the meaning of what seemed to them self-explanatory in the simple word *is*, especially when used in an ear-catching, and for them mind-boggling, repetition.

The "impurities" of language, as said before, also "contaminate" all contemporary disciplinary discourse. Even scientists sometimes feel obliged to appropriate or coin terms to describe new discoveries. Murray Gell-Mann, who discovered a new class of subatomic particles and named them "quarks," described his naming process:

I employed the sound "quork" for several weeks in 1963 before noticing "quark" in "Finnegans Wake," which I had perused from time to time since it appeared in 1939. . . . The allusion to three quarks seemed perfect. . . . I needed an excuse for retaining the pronunciation quork despite the occurrence of "Mark," "bark," "mark," and so forth in Finnegans Wake. I found that excuse by supposing that one ingredient of the line "Three quarks for Muster Mark!" [*Finnegans Wake*, 383] was a cry for three quarts for Mister. . . ." heard in H. C. Earwicker's pub.[16]

It is a measure of my veneration of the sciences—despite all apparent evidence to the contrary in this book—to take enormous satisfaction in the fact that James Joyce's imagination is enshrined in this signal scientific discovery, and to feel an extra measure of admiration for a genius in contemporary physics who is so conversant in the single most difficult book in English literature, its difficulty arising from the plurality of potential meanings in every word.

The evolving concepts of contemporary interdisciplinary theory that crosses boundaries often require borrowing (appropriating and colonizing are more pejorative terms for the same thing) the vocabulary of other disciplines, sometimes infuriating the purists among the "victimized" disciplines. One such instance occurred during our discussion of social and economic Darwinism, wherein it appeared that many biologists are scandalized at the balkanization of Darwinian verbiage, as in "economic" and "social" Darwinism: ideas they deem unrelated to pure science.

No new disciplinary area has had more to do with reforming vocabulary than what was formerly known as feminist studies, but at this evolutionary stage, might better be termed gender studies. While the word *feminism* has by now caught on with the general public as well as the discipline's detractors, its use is regarded by its contemporary practitioners as ipso facto proof of ignorance of what is currently happening in the discipline itself. One of its perennial concerns, however, remains the sex bias built into the English vocabulary; for example, the reference to everyone, regardless of sex, as "man," "mankind," or "men." This insistence on privileging the male sex in the very terminology of language had the effect of enshrining in our conscious and unconscious minds the supremacy of masculine ideals. The attack on linguistic sexual privileging was perhaps the bitterest pill for persons who were so ingrained in their own linguistic thought processes—like sexually adolescent Stephen Dedaluses—that they supposed gendered linguistic determinations to be made by God, or perhaps nature, both of which saw His/its/their views as universally acceptable.

Does science offer any analogy to or antidote for the indeterminacy of language? I think it tries to do so in its appropriation of mathematics as the irrefutable language of science, the universal, internally coherent, exact expression of results and predictability. Mathematics is an internally consistent, artificial metaphoric system conceived by man to represent reality, and believed in to the point that its presumed consistency approaches infallibility. When applied to scientific study and observation, if the numbers don't add up, the system itself can determine with equal authority the probability of error or deviation from expected results; in short, to give a verifying, partially verifying, or nugatory account of the activities to which it is applied.

Mathematics is at least as old as Western civilization itself in its Pythagorean application to the modalities of Greek music, in its positioning of heavenly bodies, and in its current application to everything from natural events to robotic precision in manufacturing. We will in a subsequent chapter discuss Plato's idea that all philosophers should first study mathematics, but I do not mean to imply here that mathematics appeared full blown and then ossified two or three thousand years ago. New fields and discoveries are as common, if not as well-rewarded, in mathematics today as they are in science, and mathematics acts as a handmaid to enable science to continue its own experimental path. But scientific discoveries often have to be explained to people like me in a language more impure than mathematics, and that endeavor plunges scientists back into the uncertainty of narrative and language and all the other flaws that flesh and the humanities are heir to.

Lest we leave this discussion of language on such a derogatory and pessimistic note, I should devote a few remarks to its positive values, most obvious in my own discipline. As old as the appreciation of literature and language itself is an intangible which we make so much of in English departments: the inherent beauty of words. The debate about the worth of literature goes at least as far back as Plato and Aristotle. For Plato the only justification for the inclusion of poetry or poets in his Republic was for the didactic purpose of using Homer's work to acquaint Greeks with their early civilization. Aristotle, on the other hand, saw poetry as something to be celebrated for its intrinsic beauty and pleasure. The debate still rages between those of us who see literature as a didactic tool of learning what is right and wrong with our society and human nature, and others for whom the sheer pleasure derived from reading and understanding exceptional literary works is justification enough. That dichotomous argument underlies the recent schisms in literature departments across the country. For me the debate implies a false choice, one that doesn't necessarily have to exist. New ways of understanding do as much to satisfy intellectual curiosity as established perspectives do to enhance our appreciation of literary works, modern as well as ancient. If we have seen the enemy, it's not us, nor should it be.

This leads to the final consideration of this discussion: the propensity of language to invoke not only accurate or imprecise information, but to evoke emotion on both the rational and subrational level. It barely needs to be reiterated that the advertising, political, and propaganda industries fashion language to their own ends. Slogans like "Now is the time for all good men to come to the aid of their country" and "Ask not what your country can do for you . . . " vie with "See the USA in your Chevrolet," "This Bud's for you," and "One investor at a time," for memorability. If such language didn't drive people to bloody war, the local showroom, taproom, or brokerage office, its politicians and advertisers would not spend their time and money to say it. While people seem to have recently gotten a bellyful of political rhetoric, they still listen and act on it as if it had a life of its own, and even the hardened can, at least occasionally, be touched by the well-turned political statement they agree with anyway.

The term "family values" is a case in point. Formerly it meant something like cheap food served in "family" restaurants. Then off the lips of politicians it became alternatively gay-bashing, right-wing ideology concerning all sorts of social problems, and finally as all-American as the Stars and Stripes. The slogan's evolution toward sanctity depended on the very imprecision of language to represent anything accurately, especially over the course of time. As a human construct, language is, like human beings themselves, flawed, mysterious, puzzling, sometimes even beautiful, but seldom scientifically accurate.

Wilson

Is the meaning of a speaker/writer always the same for a listener/reader? No. Even among scientists, can the same word have a somewhat different meaning between sender and receiver? Yes. Can the meaning of scientific words change with time? Yes. Does any of this necessarily prevent science from working? No.

A more appropriate question might be: is there enough of a shared meaning among scientists, at least some or most of the time, to allow science to work? For example, one scientist might have a different emotional response to the word "blue" than another, and even the exact range of wavelengths of light each would designate as being "blue," as opposed to, say, "violet," might differ. But the issue here is whether there is enough shared similarity in what the word "blue" means. I think the answer is obvious, since even the variation in their definitions can be reduced, if more accuracy is required, by their agreeing to an arbitrary wavelength cut-off between blue and violet, or by designing instruments to detect the color. Notice also that Zack had no actual examples of how changes in the meaning of terms *within science* have irreversibly messed things up.

There is another example of adequate communication that is much closer to home. Zack and I are in the middle of writing this book about science and literature. We come from two very different fields. Nevertheless, we share enough of the meanings of many words that we can discuss these complicated issues with each other, and hopefully make our discussions meaningful for the reader. It is true that, occasionally, differences in meaning for particular terms might throw us off, but with further discussion and analysis, it would be possible to recognize such problems and strive for more common definitions. If we did not have at least some common understandings, this book would make no sense. In the case of a group of scientists, the shared base of words with similar meanings that is necessary for their research is even greater, and so there is even less of a problem. That is part of what the education of a scientist is all about—not so much indoctrination as gaining a common understanding of terms and their definitions.

As to the issue of words changing their meanings with time, that actually is a strength rather than a weakness for science. Rather than its being necessary for science to stop any and all changes in the meanings of words in order to work, it often is the case that the meanings change, deepen, and more accurately reflect reality, as science progresses.

Take the term "gene." Its accuracy in depicting something in the real world grew with a change in its exact meaning as we progressed from Mendel's studies of inheritance, to the identification of chromosomes as carriers of genes, to the identification of DNA as the genetic material, carrying genetic information in its sequence of nucleotide base pairs. So, some changes in definition, rather than being a problem, become a strength.

There has been much more to the development of the concept of "gene." As our knowledge progressed, it was realized that the term "gene" was being used in a couple of different ways, and that this was something that required clarification. There was the idea of each gene as responsible for specifying a product in a living organism. There also was the idea of a gene as the smallest unit capable of mutation. The latter definition was found to be incompatible with the first, because parts of what we today call a gene can be mutated—individual nucleotides can change, and genes can contain hundreds of these. So, while the precise definitions of terms do develop with a science, and while different meanings can cause difficulty for a time, these can be corrected or clarified as science deepens its insights about nature. I hope this one brief example gives some idea of how problems with meaning and definitions can be dealt with in science.

But there obviously are times when changes in meaning would not be helpful. For instance, Newton published what came to be known as his laws over three hundred years ago. His laws are a mix of mathematics and narrative (equations such as $F = ma$, on the one hand, still depending upon the meaning of words such as "force," "mass," and so on, and on the other hand, brief narratives, such as "a body in motion remains in motion unless acted upon. . . ."). Since a considerable period of time has passed, we can ask if the meanings of Newton's words have changed so much that scientific accuracy is impossible, as some might infer from Zack's comments. On the contrary, we can read Newton's original words, and they still make sense to us, even if there have been some changes in meaning: the meaning has not changed so much that we may fail to grasp Newton's essential ideas. Furthermore, we are able to look back through three hundred years of scientific literature and see how his laws have been applied. In addition, the laws have become the basis of whole, applied fields within engineering. At least three hundred years have not been enough to cause significant problems for us.

In the end, what would seem more important is whether the meaning of the terms has been constant enough for science to work. There is a problem for science only if the difference of meaning over time, or among individuals, is great enough to foul up the science. So far so good, it appears.

It is not my intent to dismiss Zack's concerns totally. Problems with the meaning of terms could have been a big problem, but in practice these do not seem to have arisen. Nevertheless, we must be constantly on guard against language problems and limitations. Scientists do spend time working on definitions and meanings as disciplines and ideas develop. Fortunately, not only are scientists, themselves, directly involved in trying to avoid or fix such problems, but there are philosophers who spend their working lives on these issues as well.

Bowen

For once I was not being querulous but merely discursive in my discussion of the uncertainty of language. I didn't cite any examples of sloppy science resulting from linguistic indeterminacy or anything like that, simply because I don't know of any and the idea never occurred to me. I agree that verbal evolution or relationships are no more detrimental to science than to any other kind of human behavior. The difference between my intent and the inference drawn is a primary example of the indeterminacy I was addressing. This minute problem in communication may reflect on the very nature of our thinking, I believe. Scientists like David are committed to precision—exactitude of methodology and reliability of answers, but literary folks like me are delighted by the possibilities that indeterminacy allows—puns, paradox, irony, ambiguity, hidden meanings, and the like. But that doesn't mean I think my toaster will short-circuit because of scientific ineptitude or inability to communicate meaningfully.

Preparing for Pynchon

Thermodynamics, Maxwell's Demon, Information, and Meaning

Wilson

One purpose of this chapter is to give some background on the scientific ideas that appear to be at the foundation of Pynchon's *The Crying of Lot 49*. The concepts of Maxwell's demon, information, and entropy pop up frequently in Pynchon's book, and a reading of that book would be incomplete without at least some background. In addition, these laws and concepts give us additional, concrete examples that will be helpful as we continue our analysis of the interface between science and humanities.

The First Law of Thermodynamics

The first law of thermodynamics can be stated in different, roughly equivalent, ways. Here are three, which together should shed some light on this well-established physical law. Energy can be neither created nor destroyed; the energy of the universe is constant; the amount of energy in a closed system is constant.

While the amount of energy in the universe doesn't change, energy can be converted from one form to another, and most of the interesting events that occur in the universe and in life involve such changes in the form of energy. Kinetic energy of motion can change into the friction energy of heat, as occurs when we slam on the brakes in our car to avoid hitting another object. Chemical energy can be converted to electrical energy, as it is in a battery. Energy in sunlight can be converted to chemical energy in organic molecules by the process of photosynthesis, which must take place if we are to have food to eat.

The first law of thermodynamics was developed in the 1800s as scientists played around with such conversions of energy. The hypothesis that energy was conserved in such conversions grew to a theory as more and more kinds of energy were examined and the rule was found to hold for all such conversions. Finally, early in the twentieth century, matter was recognized as another form of energy. Thus, the amount of matter in the universe is not conserved, just as the amount of electrical energy in the universe is not conserved. Each can be converted to other forms of energy, such as heat.

The Second Law of Thermodynamics

Entropy is a measure of the disorder of a system. As disorder increases, so does entropy. Imagine that you have a container with two compartments. One compartment contains a gas, and the other does not. If a valve between the compartments opens, the gas will diffuse into the empty compartment. Once diffused, all of the gas molecules will not spontaneously return to the single compartment. These are the predictions of the second law of thermodynamics.

As with the first law, there are different ways of stating the second law of thermodynamics. Among these are: the entropy of the universe is increasing; in any physical event, the entropy of the universe either increases or stays constant; in extracting heat from a reservoir to do work, there will be an unavoidable energy loss; heat cannot be transferred from a cooler to a hotter body without expending energy.

The second law is recognized to be a statistical law. In the example above, there is nothing impossible about all of the molecules of a gas returning to the first container spontaneously, but the odds of their doing so are so low that one would have to wait longer than the universe is likely to exist just to see it happen once.

Another example of the spontaneous direction of events, required by the second law, might help. If you have warm and cold bodies, and you bring them into contact, the warmer body gets cooler while the cool body gets warmer. That is as predicted by the second law.

There are other ways of stating the second law, and of describing in what directions particular processes can proceed spontaneously, but I hope that this gives you some idea of what is involved. In crude terms, the second law states that if we want to cool a house on a warm day, we have to expend energy, releasing heat to the exterior. If the inside of the house is initially at 85 degrees Fahrenheit, and it is 85 degrees outside, we cannot just open windows and cool the house (although a breeze might cool us somewhat by evaporating sweat). If you really want to reduce the temperature inside, you must pay your power company to do it. Then, what happens is that the air conditioner cools the house by doing work and warming the outside. While the entropy of the room decreases as it cools, the entropy of room-plus-outside increases, and there is no violation of the second law.

If we have an organism that is organized, and it grows and reproduces, and the offspring grows, there is no violation of the second law for a similar reason: energy is supplied from the outside—from food—and the entropy of the world outside the organism increases more, with waste products, including heat, given off, than the entropy inside decreases. Living organisms do not violate the second law of thermodynamics. Even though we can get more organized, we use energy to do so and produce heat and other waste products that assure the continued increase of entropy when one takes account of all of the changes in us and our environments.

Maxwell's Demon

With what's been said as a background, what is Maxwell's demon? I ask because the demon is a fascinating imaginary creature, and because Pynchon talks about Maxwell and his demon in *The Crying of Lot 49*. Maxwell's demon is named after the physicist James Clerk Maxwell, who is most famous for devising a set of equations governing electricity and magnetism. That set of equations predicted electromagnetic radiation, and its speed, which matched that of light. So, Maxwell was a pretty bright guy who gave us a perspective that unified such various phenomena as x-rays, gamma rays, radio waves, visible light, ultraviolet light, infrared light, and microwaves, all of which are forms of electromagnetic radiation.

Maxwell also thought about the second law of thermodynamics. In 1867 he suggested what he imagined (not realized) could be a way around the second law. Consider two bodies of gas in two containers that have a common wall. Initially the two gases are at the same temperature. Now, the second law would say that one will not get hotter while the other gets cooler unless something else happens, like an air conditioner running between them. A hotter temperature would mean the gas molecules would have more kinetic energy, the energy of motion: they would be moving faster. At a given temperature there is a range of speeds among the molecules in the gas. The temperature of the gas is a measure of the average speed of the molecules.

Maxwell imagined the following possibility: put a tiny trap door in the common wall between the two gases. Whenever a fast moving molecule of gas is coming at the trap door from one side, open the trap door and let it through to the second side. On the second side, if a slower molecule is coming, open it and let it go to the first side. The trap door opener became known as Maxwell's demon. After a while the temperature will rise on the second side because it will have molecules of higher kinetic energy. That result would appear to violate the second law of thermodynamics since the demon has heated one side and cooled the other, thus reducing the overall entropy of the system without appearing to do anything else.

No one has ever shown such a demon to exist. There has been much written, through the years, in scientific papers, about Maxwell's demon. It remains a fantasy demon—so far, we have never found a way to violate the second law. The purpose of Maxwell's exercise was to show that the second law is only statistical—it is just that the odds are stacked against the trillions and zillions of molecules in a gas getting organized, in the absence of something like Maxwell's demon. Probabilities of all the faster molecules being found on one side of the container were extremely low, and consequently, there was no reasonable chance for a significant, measurable temperature differential between the two sides to develop.

Information and Computers

Recently, with the development of computer science, there has been renewed interest in Maxwell's demon. This is because it has been postulated that information is related, negatively, to entropy, as first suggested in 1929 by Leo Szilard (the same person who persuaded Einstein during World War II to write to President Roosevelt and propose that the United States develop an atomic bomb). In Szilard's view, the demon's efforts would be more complex than Maxwell had realized because it needed information about the speed of individual particles and where they were, relative to the trap door. Some have since argued that gaining the information, remembering it long enough to open the trap door, and then erasing the demon's memory, would release entropy and save the second law even were such a demon to exist. Others have questioned the very view that information is linked to entropy. There are actually a lot of complexities related to the actions of the demon, some of which appear to "save" the second law, even were such a demon to exist, but I will not go into the details.[1]

While we are on the topic of information, let me tell you just a touch more about computers and information. Most modern computers, such as our desktops and laptops, work on a system of bits of information. Each bit of information is a simple yes or no, a one or a zero, an on or an off. Leo Szilard (again!) was instrumental in defining this "bit" of information, which since has become so commonplace in computer parlance. Couple eight bits together and you generate $2^8 = 256$ options, enough to cover all of the keys on the computer keyboard. That's called one byte. You can view each bit as a little switch in the computer, in either an on or an off position, a one or a zero. By the right kind of manipulations one can get this information to appear on the computer screen. Each time you type a letter during word processing, and save it, you lock in a particular set of eight switches. Call up the words on the screen the next time you work at the computer and you get back what is stored at a particular site in the computer's hard drive—11011011 is a certain letter, 10111001 is perhaps the capital version of the same letter, and on it goes. You will see occasional references to this kind of thing, and to machines more generally, in Pynchon's novel, and you will find Maxwell and his demon surfacing in several places as well. I hope this helps set the stage for Pynchon's use of the terms.

Information Versus Meaning

I want to take a moment to look at something less related to Pynchon's novel, but relevant to other issues we need to explore, issues concerning human nature. I'd like to try to distinguish between information and meaning, or understanding. Information, today, can refer to data or words in a book or in a computer. We speak of information being stored in a computer. I want to make sure that we are on the same page, so to speak, in terms of a distinction between information and understanding.

I don't think that any computer, so far, has understanding. Computers, at least all of today's computers, do not grasp meanings. They manipulate symbols without knowing anything about what the symbols mean. We, on the other hand, along with chimps and perhaps a few other organisms, possess understanding. We have an understanding of the meaning of the information. We see a list of numbers and someone can tell us, "Oh, that's a list of student ID numbers and exam scores," or "That's a list that shows your salary, the various deductions that were taken out of it, and finally, your take-home pay." To a computer, there is no understanding of the significance of the two sets of numbers.[2]

Humans can grasp meanings, and that seems to be one of the things that sets us, and some other animals, off from the rest of the known universe. It seems to be something that we have not yet been able to give to, or develop in, a computer, except in science fiction, as with Hal in Stanley Kubrick's film *2001: A Space Odyssey*.

What is it that is special about humans and some other animals that gives them understanding—the sense of meaning which today's computers lack? We don't know, yet. Can we, or will we ever, build a computer that has such capabilities? The answers to these questions may lie in the outcome of what I believe to be a possible new revolution that is occurring in science. That revolution could shape how we view ourselves as profoundly as the Copernican revolution shocked us out of the center of the universe. The revolution involves science's attempt to understand the relationship between mind and brain—how experience, perceptions, and consciousness arise from brain activity. But I'll save anything more on this for a later chapter.

Thomas Pynchon's *The Crying of Lot 49*

Bowen

Although Thomas Pynchon graduated from Cornell in 1957 as an English major, he began his higher education in that institution's pioneering engineering physics program, with its emphasis on theory rather than practical applications. He worked as a technical writer for the Boeing Corporation in Seattle from 1960 to 1962, and his works continue to reflect his interest in physics and its application to his fiction and human behavior.

The Crying of Lot 49 is basically a West Coast mystery story, a search for answers as much to life's meaning as to a particular problem. Taking a paradigm from computer science with its binomial alternatives, described by David above, Pynchon composes a string of decisions that hinge on a series of unilateral choices multiplied exponentially along the way, as they are in computer technology. Some of these alternatives include the cause-and-effect chains that comprise history and historical reasoning as a way of deciphering the complexity of the world; the laws of energy and quantum mechanics; the process of energy conversion and entropy, and its relation to communication and information theory; and, in the face of uncertainty, the advent of paranoia as a corollary or alternative to conspiracy and/or chaos.

Oedipa Maas is the detective/pilgrim/truth-seeker who engages in a quest for self-definition as much as for answers to a world probably run by her former lover, real estate mogul/trickster Pierce Inverarity. Names frequently denote aspects of the characters' personalities. Oedipa's name indicates her search for self-knowledge via Sophocles' hero, as well as her potential stumbling blocks via Freud: her initial dependence on the men in her life for her inspiration. Pierce Inverarity's name is a combination of several aspects of her quest: his first name might well have been derived not only from phallic or incisive intellect, but also from a punning variant on the name of a scientist and philosopher mentioned in an earlier chapter for his work in semiotics, Charles Sanders Peirce, who in 1892 declared chance to be a basic factor in the universe:

> Try to verify any law of nature and you will find that the more precise your observations, the more certain they will be to show irregular departures from the law. . . . Trace their causes back far enough, and you will be forced to admit they are always due to arbitrary determinism, or chance.[1]

Charles Peirce and Ludwig Boltzman marked a beginning to quantum mechanics. But more about that later.

Pierce Inverarity's last name sounds like a combination of "verity" (or "truth"), "variety," and "rarity," all possibly transformed to negatives by the prefix "In," and possibly affirmed by the use of the prefix as a preposition ("in" denoting "inside of").

About Mucho Maas's first name there is less doubt, but connotations about his surname remain opaque. "Mass" seems to be the most likely, since Mucho always seems to be acutely aware of the big picture, no matter how far his speculations take him. Lost in a historic world of subjugated knowledge regarding the grim lives of the former owners of the trade-ins at the used car lot where he previously worked as a salesman, Mucho seems greatly affected by the acronym of his dealers' association, "N.A.D.A." (National Automobile Dealers Association)—or to an existentialist, simply *nada*. As the novel opens he remains dissatisfied in his present job—a disk jockey with an audience of teeny-boppers—because he fails to see any significance in his employment, whereas in used cars he was overwhelmed by the portents of despair and doom projected by the cars he bought and sold. He finds salvation at the end of the book in hallucinogens that allow him perceptions of fragmentary bits of the whole: individual notes and instruments out of tune in the cacophonous recordings he played at the station. Recognizing the minutiae of life, he loses the big picture.

Inverarity starts Oedipa on her search by making her executor of his vast estate, including real estate and a number of other enterprises, including interests in tobacco, aerospace technology (the Yoyodyne Corporation), and numerous other businesses. Her assigned mission of trying to fathom out the extent and coherence of Inverarity's holdings is like trying to figure out how the entire contemporary corporate world works, as well as her place in it. One of Inverarity's assets is a stamp collection, a project that involves Oedipa in history: in historical artifacts and their relationship to the present, in communication, and in postal services engaged in a venture that, like other means of mass communication such as the telephone, lends itself to monopoly. The rivalry of alternative means of postal communication becomes a metaphor for the bonding of disenfranchised groups over the last four centuries, culminating in a postal system called W.A.S.T.E., which (we learn toward the end of the quest) stands for "We Await Silent Tristero's Empire."

The acronym WASTE also concerns both indeterminacy and entropy. Lance Olsen gives a succinct explication of how the question of indeterminacy evolved in physics:

> When [Werner] Heisenberg attempted to describe the motion of wave-particles in terms of ordinary concepts like position and momentum, he found that a certain indeterminacy arose. The very act of observing a wave-packet disturbed it to such an extent that no accurate information could be gathered about it. At least one quantum of energy had to be used to make an observation, but the effect of this energy upon the particle was to disturb it in a random and uncontrollable way, so that it was impossible to correct for the disturbance. In other words, one could not know simultaneously to any desired accuracy the position and momentum of an object, no matter how good the instruments used and no matter how careful the procedures. . . . Thus, one has to make a choice between knowing either position or momentum. If one knows the position of a particle and not the momentum (i.e. mass and velocity), one cannot predict the particle's future position or momentum. Inversely, if one knows momentum and not position, it is equally impossible to know future position or momentum. (156)

In applying Heisenberg's Uncertainty Principle to all issues of causality, it became apparent that the inadequate relationship between the quantum theory's statistical universe of perception and the "real" world of causality is so flawed as to limit our ability to know and predict physical states of affairs, and opens up the speculation that nature may "at bottom be irrational and chaotic" (William Barrett as quoted by Olsen, 157).

Thus Oedipa is faced with bifurcated choices: either to know and measure the pattern of events which she so imprecisely observes in order to discern causality in existing circumstances, or to decide for the existentiality of a world of randomness and chaos. Motive, literal cause and effect, and the scene of her present meaningless existence are set in the first few sentences of the novel: "One summer afternoon Mrs. Oedipa Maas came home from a Tupperware party whose hostess had just put perhaps too much kirsch in the fondue to find that she, Oedipa, had been named executor, or she supposed executrix of the estate of one Pierce Inverarity." The Tupperware, fondue, and kirsch speak to the meaningless randomness of her life. "Oedipa stood in the living room, stared at the greenish dead eye of the TV tube, spoke the name of God, tried to feel as drunk as possible. But this did not work." If communication with God has something to do with the dead green eye of the TV tube, she seems lost indeed. If the reader somehow misses the point of describing a meaningless existence, Pynchon provides other highlights of her day. Through the rest of the afternoon she shops in downtown Kinneret-Among-the-Pines and listens to the Muzak, which begins for her with "bar 4 of the Fort Wayne Settecento Ensemble's variorum recording of the Vivaldi Kazoo Concerto, Boyd Beaver soloist."

Oedipa's husband, Mucho, hardly provides any alternative to the grim scene, with his own dour preoccupations, and neither do her advisers, Roseman, a nutty lawyer preoccupied with his jealousy of Perry Mason, or her shrink, Hilarius, a former German concentration camp doctor intent on enlisting her as a guinea pig in his experiments with hallucinatory drugs. Oedipa thinks of herself as a Rapunzel figure, trapped in a tower to be liberated by her prince (presumably Inverarity), using her own unfettered hair as a ladder of escape. This image is seconded by her reaction to a Remedios Varo painting of some captive young women or young nuns in a tower weaving a tapestry "which spilled out the slit windows and into a void, seeking hopelessly to fill the void: for all the other buildings and creatures, all the waves, ships and forests of the earth were contained in this tapestry, and the tapestry was the world." Oedipa's crying-jag response to the Varo painting indicates the depth of her clueless angst. Thus, the search that Inverarity—himself a shape-changing, many-voiced trickster figure—sets her on is undertaken as much for self-emancipation as an understanding of human existence.

As Oedipa approaches the valley of the city of San Narciso (the name presumably a pun on self-indulgence), from her high angle of vision she likens the regularity of its design to the circuit board of a transistor radio: "[T]here were to both outward patterns a hieroglyphic sense of concealed meaning, of an intent to communicate. . . . As if, on some other frequency . . . words were being spoken." Dominating the landscape is the Galactronics Division of Yoyodyne, Inc., with two sixty-foot missiles on either side of its entrance, clearly the commanding image of the whole area and of the day.

Oedipa registers in a motel named "Echo Courts," dominated by a sign on which is represented an enormous nymph with her dress continuously blown upwards by mechanical means "in constant agitation, revealing enormous vermillion-tipped breasts and long pink thighs at each flap." Clearly the sign and name point to the nymph Echo of mythology, doomed to repeat the words of others rather than her own. Her love for Narcissus caused her voice, destined only to repeat the words of others, to fall away, and she was subsequently victimized. The modern counterpart of this story and image with dress aflutter is the famous still photograph of Marilyn Monroe from *The Seven-Year Itch*.[2] In Echo Courts Oedipa is in turn seduced by the handsome lawyer, Metzger, from the firm attending to Inverarity's estate.

The Paranoids, a local rock group led by Miles, the motel manager, are introduced for a major role in helping to represent paranoid conspiracy theories and offer musically and otherwise an alternative to Inverarity's world of dark and mysterious control. In a bizarre seduction scene Metzger and Oedipa play a stripping game as she tries to guess the outcome of *Cashiered*, an old movie on TV. It was Metzger who, as child star Baby Igor, played a major role in the film about a minisub in the Dardanelles during World War II. The plot of the film and the history it represents become incoherent because the TV projectionist mixes up the reels, so the viewer witnesses scenes out of chronological order, adding to confusion. Only Metzger, who appeared in the film, has some inkling of its outcome. As the audience for this pathological vision of history seemingly without rational cause and effect, Oedipa in her increasing drunkenness slides toward inevitable and not unwanted seduction by Metzger as the Paranoids set up their electronic band out at the pool. During the course of Oedipa's putting on additional layers of clothing to protect her against being an easy conquest for Metzger, she accidentally knocks an explosive can of hair spray to the floor, and the compressed gas in the can precipitates an Armageddon as the object hurtles out of control around the bathroom. The chain of random events and energy leading to a calamitous conclusion is Pynchon's foreshadowing representation of chaos theory in action: the hair-spray can, a charged particle in the equation in which history is muddled and its future path unpredictable. As the Paranoids crescendo into an ever louder and more frenzied coda, their instruments, all plugged into the motel room electrical socket, blow out in one great climax as Oedipa and Metzger have their sexual equivalent on the bed. The chapter prefigures the whole plot of the novel: the search into a disconnected past, the seeming information originally envisaged as an orderly circuit board, the unharnessed energy charge, the resulting chaos, the paranoia of trying to make sense out of uncontrollable circumstance, and the dissolution of it all into entropic surrender.

Stamps and postal systems are prime means of communication, and through them Pynchon demonstrates the links between physical science and the world inhabited by his characters. As David has said, the first law of thermodynamics states that there is a finite amount of energy in the universe, while the second law tells us that the total entropy is continually increasing. When useful energy is transformed from one state to another a penalty is imposed: there is a loss in the amount of energy capable of conversion into work.

The corollary of this law in communication theory is that the greater the amount of useful energy poured into communication (communicating with people, that is), the less meaningful the communication will be. For instance, an advertising message requires an enormous expenditure of energy resources in terms of money, technology, and creativity to go out over the air or in print to large numbers of people, most of whom are not going to buy the product. Consequently, the enterprise generates an entropic dissolution of energy. Another example would be that the longer or more frequent the message is (political or otherwise) the fewer people will read or listen to it. Academic, business, or governmental administrative memos have the same result: the greater their frequency or their length, the more meaningless they become to their recipient. On every floor of our university building there is an enormous container for waste paper, a quixotic attempt to salvage something recyclable, hence reusable energy, from the entropic waste created mainly by the vanity of the senders. A prime example of informational entropy is, of course, the Internet, capable of transmitting exponential amounts of increasingly ignored or irrelevant information that falls for the most part on the deaf ears or eyes of all but the passionate few. Just as heat energy seeks its normative level in the undifferentiated blob, so does communication find its own level of ennui or meaninglessness. What we intend to say, what was so important for us to say, comes through skewed in both linguistic translation and meaning, as discussed before, and certainly with far less meaning than it had for the communicator. Ultimately, when the world runs out of useful energy, it correspondingly will run out of meaningful knowledge. The result is the nada that so terrified Mucho.

Ironically, then, Oedipa's potential solution to her personal identity problems—to "bring to an end her encapsulation in her tower"—may come from her discovery of the Tristero System, the enigma that haunts her throughout the novel. She begins a mini-epic search up and down the north coast of California, an effort that leads either to a return to entropy or some alternative meaning that would explain the nature of life. In a bar near Yoyodyne, Oedipa discovers the existence of an underground postal service, by which subscribers—mostly dropouts or the disaffected of one kind or another—communicate not so much for the reward of articulating meaningful news, but simply to belong to a system outside political hierarchical control. What evidence there is that such a system exists is plentiful, but fragmentary and elusive. The stamps or envelopes carry the symbol of a muted posthorn or references to WASTE.

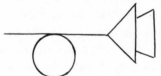

Figure 1: Pynchon's Muted Posthorn

WASTE, of course, has a lot to do with entropy. So does the enormous amount of energy Oedipa puts into the search for answers. She discovers her first connection to the system through Mike Fallopian, whose name embodies a sort of gynecological preincubation tube for the idea. Besides sending and receiving mail through WASTE, Fallopian is a member of the Peter Pinguid Society, named for a confederate naval officer who fought (or did not fight) a battle with the Russian navy in 1864 off San Francisco. The putative engagement, which resulted in no discernible outcome, "was the very first military confrontation between Russia and America," although, during the American Civil War, the Russians were supposedly allies of the North (United States) against the Southern Confederacy. The Pinguid Society exists to celebrate this non-event as a milestone in American history. Fallopian is one of those people communicating such non-messages by WASTE.

Although I don't actually know if Peter Pinguid was a real historical person, nearly every seemingly absurd historical event in Pynchon that I have checked on does have some basis in historical fact. For instance I had just read *Lot 49* for the first time, and was at a conference in Copenhagen when I chanced to walk by a post office and was startled to find the posthorn sign swinging over the door. It was what I now take to be the coat of arms of the Thurn and Taxis, the family-run postal empire of Europe dating back to 1300. The variant of adding a mute to the horn would then represent the effort to silence or mute the postal communication by the Tristero, who chose the muted horn as their logo. The sight of the posthorn itself was as much a revelation to me as to Oedipa, like witnessing a bleeding statue of the B.V.M. No experienced Pynchon scholar whose work I have read ever rejects out of hand any of the author's historical assertions, no matter how unlikely (for example, the alligators in the New York City sewer system in *V*).

Fallopian, engaged in writing a historical account of the U.S. mail, details the nineteenth-century laws enacted to suppress competition with the government's postal monopoly, an action that Fallopian sees as a parable of the systematic abuse of power. With this out-of-sequence example from nineteenth-century America Oedipa begins her own reconstruction of the history of mail systems in Western Europe and of the Trystero organization's migration to the United States. ("Tristero" is used interchangeably with its historical antecedent "Trystero" here and in Pynchon's text.)

During a trip to lake at "Fangoso Lagoons" (an Inverarity housing development) into which Pierce had put human bones as artifacts for diving enthusiasts, Oedipa discovers the bones' possible connection to the remains of a company of American soldiers who were slaughtered by the Nazis during World War II at Lago di Pietà in Italy—the whole episode tied into the mail conspiracy by a similar slaughter perpetrated on pony express riders by the Tristero in 1853 in their long conflict with rival mail companies. While the origin of the postal-empire conflict unfolds gradually throughout most of the book, the answer to its relationship to a widespread contemporary conspiracy is never fully defined.

The historical mystery begins to unravel when Oedipa attends a local production of Richard Wharfinger's seventeenth-century melodrama, *The Courier's Tragedy*, directed by Randolph Driblette. The fourth act of the play ends with the lines,

> He that we last as Thurn and Taxis knew
> Now recks no lord but the stiletto Thorn,
> And Tacit lies the gold once-knotted horn.
> No hallowed skein of stars can ward, I trow
> Who's once been set his tryst with Trystero. (75)

The mention of the historical Trystero as well as the rival monopoly, Thurn and Taxis, indicates that their earlier back-and-forth bloody conflicts for the European postal trade may have eventually been imported into the United States by immigrants still harboring murderous loyalties to the Trystero. Clues link the three lakeside massacres—the one in the *Courier's Tragedy*, another in World War II, and the pony express massacre. The most explicit is the perhaps senile testimony of old Mr. Thoth (a variant on the name of an ancient muse) regarding his grandfather's eyewitness account of how the black murderous figures of the pony express riders' assassins were not Indians as claimed, but the Trystero. In sum the stories form the basis of possibility for a genuine chain of historical events linking the contemporary WASTE system to an ancient but ongoing multinational conspiracy.

Oedipa's search focuses on the suppression of the play's Trystero reference and its momentary reemergence in the Driblette production. This takes her to an antiquarian book dealer, a Berkeley textual scholar, and a stamp dealer, all of whom offer increasingly tantalizing clues on the censorship and variants of the lines, as literary/scholarly answers are sought for the historical power struggle over the control of communication.

Meanwhile, on the scientific front, new technical information and theory is forthcoming from Oedipa's visit to Yoyodyne and her meeting with Stanley Koteks, for whose surname I offer only a partial gloss. In addition to the obvious, the name Koteks suggests someone associated with corporate *tech*nology. In the following scene Pynchon directly relates particle theory to Oedipa's search. Koteks doesn't agree with his boss, Bloody Chiclitz (pun intended), who is always harping on team effort (co-techs). Koteks offers an example of individual initiative in John Nefastis's invention, the "Nefastis Machine" which ironically requires cooperative effort to operate:

> From a drawer he produced a Xeroxed wad of papers, showing a box with a sketch of a bearded Victorian on its outside, and coming out of the top two pistons attached to a crankshaft and flywheel.
> "Who's that with the beard?" asked Oedipa. James Clerk Maxwell, explained Koteks, a famous Scotch scientist who had once postulated a tiny intelligence, known as Maxwell's Demon. The Demon could sit in a box among air molecules that were moving at all different random speeds, and sort out the fast molecules from the slow ones. Fast molecules have more energy than slow ones. Concentrate enough of them in one place and you have a region of high temperature. You can then use the difference in temperature between this hot region of the box and any cooler region, to drive a heat engine. Since the Demon only sat and sorted, you wouldn't have put any real work into the system. So you would be violating the Second Law of Thermodynamics, getting something for nothing, causing perpetual motion. (85–86)

Oedipa, like her scientific predecessors, is quick to point out the flaw in Maxwell's scheme.

> "Sorting isn't work?" Oedipa said. "Tell them down at the post office, you'll find yourself in a mailbag headed for Fairbanks, Alaska, without even a FRAGILE sticker going for you."
> "It's mental work," Koteks said, "But not work in the thermodynamic sense." He went on to tell how the Nefastis Machine contained an honest-to-God Maxwell's Demon. All you had to do was stare at the photo of Clerk Maxwell, and concentrate on which cylinder, right or left, you wanted the demon to raise the temperature in. The air would expand and push a piston. The familiar Society for the Propagation of Christian Knowledge photo, showing Maxwell in right profile, seemed to work best. (86)

It is difficult to place an interpretation on the significance of Nefastis's experiment: it is so off the wall in relation to what Maxwell conceived as a metaphoric representation of thermodynamics. But its fame among scientists lay in its literal possibilities, which they debated for years, as if metaphor and literal truth were equal. But in *Lot 49*, which always walks the line between fantasy and reality, the extension of Maxwell's Demon to Nefastis's machine might very possibly be a warning not to take all of Pynchon's playful metaphors too literally. Oedipa does visit Nefastis and tries to focus her concentration on Maxwell to the point of moving a cylinder. She fails, of course, just as she is bound to do in her expenditure of inordinate amounts of energy in ultimately proving the unprovable conspiracy theory. At the end of the book she may be on the verge of identifying a member of the Trystero organization, and, on the other hand, she may be as far from it as she ever was. That is in a sense Pynchon's brand of realism. It doesn't matter about finding the ultimate truth so much as imagining it and searching for it. If Oedipa does not move the cylinder in Nefastis's machine by the power of her mental concentration alone, she does learn about the affinity between the entropy of thermodynamics and the entropy of communication theory—the principal difference between Maxwell's box and Nefastis's invention:

He began then, bewilderingly, to talk about something called entropy. The word bothered him as much as "Trystero" bothered Oedipa. But it was too technical for her. She did gather that there were two distinct kinds of this entropy. One having to do with heat-engines, the other to do with communication. The equation for one, back in the '30's, had looked very like the equation for the other. It was a coincidence. The two fields were entirely disconnected, except at one point: Maxwell's Demon. As the Demon sat and sorted his molecules into hot and cold, the system was said to lose entropy. But somehow the loss was offset by the information the Demon gained about what molecules were where.

"Communication is the key," cried Nefastis. "The Demon passes his data on to the sensitive and the sensitive must reply in kind. There are untold billions of molecules in that box. The Demon collects data on each and everyone. At some deep psychic level he must get through. The sensitive must receive that staggering set of energies, and feed back something like the quantity of information. To keep it all cycling. On the secular level all we can see is one piston, hopefully moving. One little movement, against all that massive complex of information, destroyed over and over with each power stroke." (105–6)

It is hard to read this passage closely without sensing the parallel between Nefastis's creation and Pynchon's in *Lot 49:* Inverarity, as the Demon, passes his data on to the sensitive Oedipa, in hundreds of fragments. Inverarity collects (has collected) data on each and every one. He is trying to get through "at some deep psychic level." Oedipa receives his staggering set of energies, and feeds back something like, or perhaps less than, the same quantity of information, having sorted it into meaningful patterns that lead on to other potential patterns "To keep it all cycling."

Having offered us a major clue as to his plot, Pynchon suggests his reasons in terms of metaphor:

"Entropy is a figure of speech, then," sighed Nefastis, "a metaphor. It connects the world of thermodynamics to the world of information flow. The machine uses both. The Demon makes the metaphor not only verbally graceful, but also objectively true."

"But what," she felt like some kind of a heretic, "if the Demon exists only because the two equations look alike? Because of the metaphor?"

Nefastis smiled; impenetrable, calm, a believer. "He existed for Clerk Maxwell long before the days of the metaphor."

Here Pynchon comes close to revealing himself as a self-reflexive novelist. The author has created Inverarity as Maxwell's Demon, a metaphor—shadowy, graceful, but objectively true in terms of what shape-changing, trickster figures are. The conversion of objective reality into metaphor that we have been discussing throughout the book is a result of our need to express some sort of intuitive reality. While it may not drive a piston, it provides the energy of human intuition, contemplation, and communication. Nefastis is, then, a surrogate of the author himself, working out of a Clerk Maxwell–like belief and experiment in metaphor. The characters are metaphors for ideas and those ideas/metaphors themselves breed other metaphors by which to bring the sensitive—Oedipa—to realizations which she, as a fictive metaphor herself, passes on to the next level of sensitive—the readers. We accompany her on her search; our reading has to many of us seemed excessively energy-consuming; and our understanding in the form of moving piston/brain cells probably has yet to occur. When Oedipa doesn't get the machine's piston to move she comes to the same conclusion that many of this discussion's readers might be expected to have: "The true sensitive is the one that can share in the man's hallucinations, that's all."

Oedipa, supercharged with both scientific theory and literary history, now begins to examine more intuitively the world around her. How does the conspiracy, if any, affect contemporary life? Where clues at first were difficult to come by, she enters into a night of almost surreal half-intoxicated wandering, gathering all sorts of random information—stories and characters who all confirm the existence of the underground mail service. For the most part they are a nondescript group of victims of one sort or another. This long journey through all-night San Francisco is in the literary tradition of the *Walpurgis Nacht,* a surreal journey of exploration somewhere near the unconscious or dark side of the mind made by Goethe's *Faust.* It constitutes a revelatory mixture of everyday reality and our mental impressions, conscious and unconscious. A few examples that might be familiar are the restaurant scene in *The Death of a Salesman,* where present reality mingles with Willy's guilty memories of infidelity, and *Who's Afraid of Virginia Woolf?* where during a long drunken night George and Martha work through the fantasies of their tormented existence. Most representations are exhausting ordeals for the participating characters, just as Oedipa's dark night of the soul brings her to a point near despair. Returning to her hotel, she encounters a convention of deaf mutes dancing wildly varying steps to soundless music, like particles in wild motion but never colliding as if all participate in "a choreography in which each couple meshed easy, predestined." Randomness, predestination, chaos, and despair form the nadir of Oedipa's search.

Oedipa returns to Kinneret to find Hilarius a guilt-ridden, raving lunatic, and Mucho with his mind blown on Hilarius's hallucinogens. Clearly she can't go back home again. Back in San Narciso, she learns that Metzger has flown the coop with one of the Paranoids' teeny-boppers, and now on her own she encounters a whole new series of historical complications and affinities with other aspects of the unfolding conspiracy story, as its historical relation to the scientific and political and psychoanalytic chains of events begin to merge. Wharfinger's play, *The Courier's Tragedy,* is entwined with the Scurvhamites, an ultrapure sect of Puritans, hung up on predestination, like Oedipa at the end of her collisionless deaf-mute dance. Again the Scurvhamite philosophy assumes a bifurcated separation.

There were two kings, Nothing for a Scurvhamite ever happened by accident, Creation was a vast, intricate machine. But one part of it, the Scurvhamite part, ran off the will of God, its prime mover. The rest ran off some opposite Principle, something blind, soulless; a brute automatism that led to certain death. (155)

The mention of "machine" recalls Nefastis's machine, run either by God or the devil—divine predestination or evil chaos.

Driblette, the director of the play, walks into the ocean, his intended message, if any, obscured. Oedipa concentrates on him as she did Maxwell's face, "But as with Maxwell's Demon, so now. Either she could not communicate, or he did not exist." The bookstore burns, and Oedipa's sense of paranoia grows.

Either you . . . have stumbled . . . onto a network by which X number of Americans are truly communicating whilst reserving their lies, recitations of routine, and betrayals of spiritual poverty, for the official government delivery system, maybe even unto a real alternative to the exitlessness, the absence of surprise to life, that harrows the head of everybody American you know, and you too, sweetie. Or you are hallucinating it. Or a plot [by Inverarity] has been mounted against you so elaborate and expensive . . . so labyrinthine that it must have meaning beyond just a practical joke. Or you are fantasying some such plot, in which case you are a nut, Oedipa, out of your skull. (171)

Oedipa "had dedicated herself . . . to making sense of what Inverarity had left behind, never suspecting that his legacy was America."

At the end, the answer to the whole Trystero conspiracy—whether there is indeed a conspiracy—depends on who shows up to buy Lot 49 of Inverarity's collection, the "Trystero 'forgeries,'" as Oedipa settles back waiting for the announcement, or "crying," of Lot 49.

9

Perception and Reality

Part One: Human Perception

Bowen

Historically, one of the most significant documents having to do with human perception is Plato's "The Allegory of the Cave," chapter 25 of *The Republic of Plato,*[1] a book about a utopian state ruled by a philosopher-king. As its name suggests, the chapter is a wholly imaginative narrative conveying an abstract or spiritual meaning behind concrete, material forms. This, indeed, is the very aim of a book about the supremacy of pure thought, or in Plato's term, *knowledge,* written in an era before that of modern science with its basis on close observation of the natural world and experiments involving it. Plato's idea was that an understanding of reality depended on training in mathematics and philosophical reasoning to condition our understanding of the perishable and ever-changing phenomena in the sensible world. An allegory, conceived in reason and imagination, then, demonstrated knowledge in a purer form than did untutored observation based solely on our human senses.

Rooted in this allegorical abstraction of ways of acquiring knowledge, it was nearly axiomatic that a number of various interpretations would arise over time to satisfy a variety of philosophical and religious arguments. I'd like to begin with a simplistic interpretation that most Platonists can agree on, and then offer a couple of the many varieties of religious and secular interpretations that ensued.

Plato tells us:

Imagine the condition of men living in a sort of cavernous chamber underground, with an entrance open to the light and a long passage all down the cave. Here they have been from childhood, chained by the leg, and also by the neck, so that they cannot move and can see only what is in front of them, because the chains will not let them turn their heads. At some distance higher up is the light of a fire burning behind them; and between the prisoners and the fire is a track with a parapet built along it, like the screen at a puppet-show, which hides the performers while they show their puppets over the top. . . .

Now behind this parapet imagine persons carrying along various artificial objects, including figures of men and animals in wood or stone or other materials, which project above the parapet. Naturally, some of these persons will be talking, others silent. . . . Prisoners so confined would have seen nothing of themselves or of one another, except the shadows thrown by the fire-light on the wall of the Cave facing them. . . .

Now, if they [the prisoners] could talk to one another, would they not suppose that their words referred only to those passing shadows which they saw? . . .

And suppose their prison had an echo from the wall facing them? When one of the people crossing behind them spoke, they could only suppose that the sound came from the shadow passing before their eyes. (227–28)

To recapitulate, since we are all "captives" in an environmentally controlled cave, unable to benefit directly from the light of truth and our perspectives shackled by chains so that we can only base the "truth" of what we see on our perception of the shadows on the opposite cave wall, we will think that whatever shadowy figures appear in front of us are reality. The shadows are made by passing figures or images in front of a fire, all happening behind us and out of our sight, since our shackles confine our gaze only to the opposite wall in front of us. If this spectacle of shadows is all that we ever see, we think that that is the only reality. If we can't see the other prisoners we would suppose that even what they say, bounced off the opposite cave wall, to be the speech of those shadows passing in front of us.

Further, if a prisoner were to escape and make his way out of the cave into the sunlight he would be blinded by the light for a time until finally he would become acclimated and be able to make out the sun and those things it illuminated. Returning to the cave, he would be laughed at by the other prisoners he was trying to free from their ignorance and eventually killed, as was Socrates, who was only trying to bring the illumination of truth to his fellow Athenians by his instruction, culminating in his remarks at his trial, condemnation, and execution. The allegory makes a point of the obligation of those possessing knowledge of the light (essential truth) behind the world of appearances (shadow figures) to try to educate their fellow citizens and former prisoners.

In casting the prevalent human condition as that of a group of prisoners manipulated by the preconceptions of the politically powerful who hold sway over us, enslaving us by their will and obstructing us from truth, Plato's allegory, at least in this respect, anticipates Foucault. The colonial powers, brandishing their Bibles as the all-consuming truth, had a convenient rationalization for their conquests, but no monopoly on truth or knowledge.[2]

On the other hand, religiously minded Neoplatonists have linked the nonmateriality of truth to some essential mystical experience: the source of the light with a monotheistic God, and the observation of reality as conditioned less by mathematics than by the religious dogmas of the church ("For now we see as through a glass, darkly, But then face to face"). Dante seized on the concept of equating God with the "light" of Roman Catholic Christianity in the *Divine Comedy*, when, at the end of Dante's travels as a live human being through the enormous crater/cave of hell, where sinners eternally live out the misperceptions of their lives, through the ethereal instruction of Purgatory, and through the celestial bliss of heaven—all manifested in complex mathematics—he beholds God as a point of pure light.

Philosophical Platonists like Berkeley and Hume have used Plato's allegory to envision a world wholly outside of our sensible perceptions ("We think; therefore we are"), a world completely nonmaterial, created in our individual minds. It gives rise to such old conundrums as the question, "Does a tree fall in the forest if no one sees it or hears it?" If an accident in the street has three witnesses, often each will produce a different version of what happened in terms of sequence, cause and effect, and ultimate meaning.

Indeed the idea and what it suggests have major applications for humanistic study even today. In contemporary literary criticism a widely accepted critical school called "Reader Response" deals with the cooperation of reader, writer, and the intervening text, with the final interpretation of meaning falling heavily on the reader, whose ultimate interpretation of the work is its last iteration.

When the National Football League began to resort to instant television replay in close decisions, it was based on the idea that there was some reality apart from the field umpire's perception, but although the time-honored photo finish of horse races is the official word and some races result in a dead heat, a close football call has to go one way or another. Even if we decide on the reality of the occurrence, its interpretation is hardly of lesser importance.

Any incident in the national or international news produces a chain of events and their interpretation that varies from country to country, or ideology to ideology, completely at odds with each other even when the specifics of the event are agreed on. When a boatload of Cuban refugees capsized off the coast of South Florida, the event provoked profound disagreement among the Cuban exile community, the Cuban government, and the United States. One survivor, a preschool boy, became for the Cuban government a victim of kidnaping to be returned immediately to his father, a devotee of Castro. The Cuban exile community in Miami saw him (via the image of his drowned mother) as a courageous freedom seeker who should not be ripped from the freedom-loving country of his choice, while the United States government, caught in an election year between an ultraconservative anticommunist group and the increasingly bellicose response of a sovereign country ninety miles away, tried to sort it all out politically by appealing to the federal judicial system for a ruling. National public opinion polls indicated a preference for the return of Elian Gonzalez, but whether or not public opinion dictated the course of law and eventually government action and policy remains a matter of conjecture. For our purposes the point here is that while the events of the case remain externally verifiable, their interpretation is nearly always up for grabs. That is why we have a legal system with judges and juries.

In another example, one of my academic colleagues, a blind psychologist, was concurrently a psychotherapist at a western New York State mental institution. He once told me that a schizophrenic he been working with for months had almost convinced him that the schizoid world the patient claimed to inhabit made more sense than the world we lived in. The paranoid world we read about in Pynchon is not without its own claims to truth. Academics, who probably do more thinking about perception and reality than most people, might agree with my psychologist colleague when it comes to their own particular institutions, while ad agencies celebrate the distortion of reality in their business. Political perceptions are so skewed that one newscast from a network owned by a conservative corporate conglomerate takes forty-five seconds daily for what they call a "reality check," as if an additional minute or so of biased interpretation will ultimately reveal "reality."

In a sense, we look to science, social science, and, yes, sometimes even the humanities, to ascertain something, anything, we might perceive to be truth, but every "discipline" by definition builds its own cave with its own procedural blinders, its own shibboleths, its own exclusions, all of which are in a sense the mind-forged manacles or chains causing us to perceive only the designated shadowy "disciplinary" imitations of reality on the opposite wall of the cave, and to bar any other rays of light and inspiration from the world outside. Perhaps the apparent success of science owes much to its rigorous exclusion of the dross that doesn't fit its narrow parameters of scientific methodology. That is why David can so succinctly describe those tenets, while I am left to garner by example the amorphous impressions of what humanistic thinking is all about. The more rigor our particular "discipline" displays the narrower our conditioning and the more idiosyncratic our perception. But Plato thought it was worthwhile to attain a worldly nirvana of pure truth, and some of us, like my friend David, have not completely given up that hope, and I suspect that is a major factor in our thinking it worth the while to write this book.

Part Two: Reality and Representation

Wilson

An underlying theme of Pynchon's book was the questioning of the nature of reality, and views of both reality and mistaken reality abound in literature. Perhaps it is time for us to look more carefully at what science can or cannot tell us about reality. What is it? What's really out there in the world? We each have our experiences. Does the external world exist without us to experience it? I will make what I think is the obvious answer to such questions—there is an external world that exists. I exist, and you exist, as parts of that world. There are other views, and perhaps Dr. Bowen will champion one of them, but I think that most are philosophical speculations. While there is no absolute proof possible here, there may be an understanding why some of us, at times, get bored with aspects of philosophy. I accept that you exist, independently of me, even if you think that I am a figment of your imagination. I also accept that there is a universe, that I am a part of this universe, and that the universe existed long before I, or any other human, existed.

The issues that I would like to address include: how direct is our knowledge of the world? What is the world *really* like? What can we or do we know about it? Are there limitations imposed on our knowledge of it because of our nature—because we are humans?

Indirectness

How directly do we know the external world? It seems obvious at first that we can see it, so why am I asking? The answer comes from our current understanding of what is happening when we see something, or hear something, or feel something. We *appear* to see something quite directly, but we actually have only indirect input.

As Zack has pointed out, using Plato's allegory of the cave, what we really see are shadows of the world. We gain information about the external world from our sensory inputs—the eyes for vision, ears for hearing, and the rest. That information is indirect, as described below, and it also is a narrow window through which we gain sensory input about the world outside.

Representation

The eye acts like a lens to focus an image on the retina at the back of the eye. If you are under the assumption that we are simply seeing that image directly, since that is what seems to be happening as we "look outward" from our eyes, consider the fact that the visual image on the retina is upside down and backward! Just as with a camera lens, what is "up" in the world is "down" on the retina. When first realized centuries ago, this simple observation gave old-time philosophers quite a worry. They puzzled over how the image got "right-side-up." But there is no real worry because there is no simple image that is going from the eye to the brain. Were there to be such an image, one might think that there would need to be a "little person," a homunculus, inside the brain who would view it. Sometimes that is how we see ourselves, but it won't work as a full explanation. Otherwise, there would have to be an even smaller homunculus inside the little brain of the first one so it could see for the little homunculus, and as this sequence of ever-smaller homunculi keeps going, we would have an infinite regression.

Instead, what we see is a reconstruction, a representation, and not reality directly. We have receptor cells in the retina that are sensitive to the light. Through a complex set of reactions, the light energy triggers electrical signals in retinal nerve cells. After processing through several layers of such nerve cells, a final set sends electrical signals back to the brain. We actively create our visual perception, our biased representation of reality, on the basis of that sensory input and brain processing.

This is true of our entire sensory input—vision, taste, hearing, smelling, and body sensations such as touch. What reaches the brain from all of these are electrical impulses in neurons. Each of the nerve cells can carry so-called nerve impulses—voltage signals that propagate along the neuron's fibers, or axons. In the optic nerve, there are hundreds of thousands of such nerve-cell axons that transmit the electrical signals from the eye to the brain.

In the case of visual perception, bursts of electrical impulses in the optic nerve stimulate brain neurons in the lateral geniculate nucleus, deep in the brain. The lateral geniculate neurons transmit their version of the sensory information to the rear of the human cerebral cortex, where the analysis of visual information gets serious. The input is digested and then sent on to different parts of the cortex for further simultaneous analysis—motion in one place, color in another, nature of the object in another, and so on. By combining the sensory input with memories, we come to identify objects, and infer positions and movements. Our analysis is hardly just a simple reconstruction. We emphasize contrast, or differences, between parts of our visual image, and this probably helps us to distinguish objects from background. We also actively pick and choose what we want to pay attention to.

Reality

Can such indirect input and analysis, such shadows, ever allow us to gain a sense of what reality is all about? Are we forever limited by our sensory inputs, unable to correct for any misperceptions that we have? In the case of vision, there is good evidence that we can become aware of errors and misperceptions. Consider optical illusions and ambiguous figures. As described so well in Gregory's *Eye and Brain,* we are susceptible to a number of illusions, but it is clear that we have been able to recognize them, and so to correct for them as we develop a science of vision. Ambiguous figures arise because two-dimensional drawings can sometimes have more than one possible three-dimensional counterpart. You may have seen such ambiguous figures before—an image that can be an old or young woman, or an image that can be viewed as a rabbit or a duck. Perhaps the simplest of these consists of a simple line drawing of the edges of a cube, viewed at an angle (fig. 2). This is a Necker cube, which can be perceived as either one of two different cubes, depending upon which face of the cube is considered to be facing outward from the page. One's perception, or mental image, actually can shift from one cube to the other. Nothing is changing in a drawing of the cube (the environment) as the shift occurs. Instead, it is the processing going on in our brain that changes. With such ambiguous figures we come to recognize the indirectness of our visual perception, and its reliance on brain processing.

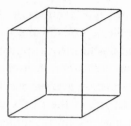

Figure 2: Necker Cube

There are other ways that we are limited by our sensory inputs. Our senses do not capture the full complexity of phenomena in the external world. For instance, in vision we can see only a limited range of wavelengths of light. We see from about 400 nanometers (purple/blue) to about 700 nanometers (red). It is clear that this has not limited us from discovering light with other wavelengths, such as ultraviolet and infrared light. We can't see these, nor can we see x-rays, radio waves, or microwaves, but we have learned about these, thanks to Clerk Maxwell, as we have discussed. And we can detect these other electromagnetic radiations with instruments and make use of them to look at broken bones, listen to music on the radio, or heat our dinner. Thus, we have developed extensions of our senses that allow us to detect phenomena in the world of which we have no direct awareness. So, it is clear that the narrow window through which we sense the external world has not limited us from learning more about that world than we can perceive directly.

We have advanced beyond our initial axioms and assumptions about the world, and beyond at least some of our sensory limitations. We can recognize the indirectness of our sensory input and we can correct for illusions. Thus, we can overcome at least some of the mistakes and limitations in our sensory and perceptual systems. Furthermore, we have been able to catagorize such seemingly different things as x-rays, radio waves, and visible light into a single kind of thing—electromagnetic radiation, as Maxwell first realized.

In even more fundamental ways we also have been able to alter our understanding of matter. How solid are tables and chairs? To our perceptions, they certainly appear quite solid. We sit on chairs and do not worry about falling through. But we also have come to understand that chairs and tables are made of atoms, which, individually, are not so uniformly solid. We have come to understand how, while our fist cannot pass through a table, gamma rays can. On the basis of experiments and atomic and quantum theory, we can understand properties of tables that seem to contradict our initial sensory view of their solidity.

So, we know that we can stretch beyond some obvious limits that might have existed on our ability to know what reality is all about. Can we go all the way? How close can we get to reality, to truth? That's still an open question. What is clear is that at least some of the (incorrect) axioms and assumptions that we "naturally" made as humans, and that were the basis, or starting points, for the development of science did not limit us: we have been able to build more accurate assumptions and consequently more accurate and complete scientific views of the external world.

We will deal in more detail with the modifications in our assumptions and axioms in the next chapter when we consider the major revolutions that have occurred as science developed. This analysis will also lead us toward some tentative answers to the question of what progress in science is all about. Some have questioned the possibility of such progress, but I think that their concerns are exaggerations. Progress may not be as simple as might have once been thought, and it is not necessary that every step in science be in the direction of truth for there to be progress in the long term. We can have detours and dead ends, but the ultimate question is whether we are generally heading in the right direction. I have few doubts about that, as you will see. As we discuss revolutions, I also will claim that we could be in the middle of a very far-reaching one that could impact on the humanities as well as the social sciences.

10

Revolutions in Science

Wilson

As we continue our exploration of the nature of reality and the question of whether truth can be approached by science, I will describe four major revolutions that have occurred in science, and suggest that we may be at the beginning of a fifth. Each of these revolutions has had a profound impact on how we think about ourselves and our place in the world, and so have had major impact on the humanities as well.

At least two of these are accepted as revolutionary in another sense, that of radically changing the foundations of science. These two are the kinds of scientific revolutions that Kuhn describes in his book *The Structure of Scientific Revolutions*, which we discussed in chapter 2. Especially relevant to our considerations in this book are the statements of some historians and sociologists who argue that the existence of such scientific revolutions, and the possibility of new ones, implies that there is no progress in science—no progress toward truth or even toward increasing accuracy. I disagree and will tell you why as we explore these issues.

I also want to raise the possibility that we are beginning a fifth revolution that will have a major impact on the humanities and our views of human nature.

Revolution 1. Copernicus, Galileo, and the Place of Earth in the Universe

Most educated people are aware of this revolution, which was mentioned briefly in chapter 2. We are not at the center of the universe, with the sun and stars moving around us. Instead, our sun is but one star among billions in the outer reaches of the Milky Way galaxy, itself but one of billions of galaxies in the universe. Our earth joins Mercury, Venus, Mars, and the other planets in elliptical orbits around our much more massive sun. Most feel quite comfortable with this realization today, but it was a radical reorientation hundreds of years ago, when humans believed that the universe was created just for them, and that they were necessarily at the center of it. Looking up from today's scientific perspective at the vastness of the heavens, the concern seems more that we shrink to insignificance.

Revolution 2. Modern Physics: Quantum Mechanics and Relativity Theory

We have already considered the impact of quantum mechanics—the wave theory of matter, the uncertainty principle, and so on, in our earlier discussion of determinism and free will. This part of the second revolution, which began about a hundred years ago, forever changed our view of the nature of matter. The altered paradigm has caused a rethinking of determinism and uncertainty.

The other half of the modern physics revolution began in the mind of Albert Einstein, one of the greatest scientists who ever lived. His relativity theory contributed its own share of shockers. Time is relative and, in some ways, like a fourth dimension with space. Euclidian geometry, which we all grow up with in school, is only an approximation, and is not the geometry of the real world. Instead, space is curved, especially near high concentrations of mass, that is, in high gravitational forces.

As with quantum mechanics, relativity theory leads to some results that contradict common sense, and again point to our ability to go beyond our seemingly innate views about space and time. One famous example is the so-called Twin Paradox. Imagine identical twins on Earth. One goes aboard a rocket-powered space vehicle and accelerates away, reaching speeds close to the speed of light (186,000 miles/second), maintains roughly that speed for, say, ten years, while taking a trajectory that returns him to earth. Now ten years older, he steps out of his space craft to greet his twin brother, who is thirty years older than when the twin left! How is this possible? Time is relative, and, as measured from the twin on earth, has passed more slowly for the traveling twin.

There is good experimental evidence that this paradoxical result, of time passing at different rates in different frames of reference, is correct for anyone who accelerates to a speed near the speed of light and stays at that speed for a length of time. Indeed, the closer the individual gets to the speed of light, as measured by someone back on earth, the slower would time pass for the moving person relative to the one back on earth. Some of the evidence favoring this view comes from observing the life-times of elementary particles, which we can get to move at nearly the speed of light in accelerators. The faster the particles are moving, the longer the apparent lifetimes that we measure for them from our per-spective.

Not only does the relative rate of the passage of time change with speeds near that of light, but, if we measure the mass and length (in the direction of motion) of objects moving at high speeds, we see that mass increases and length decreases as they approach the speed of light. All these measurements—time, mass, and length—become relative, depen-dent on the speed at which something is moving.

The modern physics revolution has had profound effects on how we view the world and matter, and has changed us in other ways. The realiza-tion that matter is another form of energy has led to the new technolo-gies of atomic and hydrogen bombs and nuclear power plants. Beyond this, our better understanding of the nature of matter may contribute, some day, to our understanding of the nature of mind and consciousness.

Progress in Science

The two scientific revolutions we have just considered profoundly changed some of the most fundamental assumptions and axioms that underlie much of science. Given that such radical changes in viewpoint can occur, some people have concluded that we cannot make progress in science, or, at least, that we cannot know what progress really is. They believe that the paradigm shifts that came with these revolutions have changed the nature of the questions that scientists can properly ask. The very meaning of terms change. "Time" did not have the same meaning to Einstein as it had to Newton. So how can we say that Einstein's view is better? Is it just different?

My response: with each of these revolutions, indeed, the meaning of terms has changed. They have been refined. I would argue that scientific revolutions have allowed us to get closer to reality. That is part of the way that science progresses—by refining the meaning of terms, making them fit reality more closely, rather than our preconceptions of it.

This kind of change in meanings for terms occurs in science even without scientific revolutions. For instance, as was discussed earlier, the idea of a gene started as an abstract thing, but it has become quite concrete with DNA, and we now can talk about alleles, mutations, the RNA and protein products of genes, the genetic code, and so on. Has the meaning of "gene" changed in the last hundred years? Yes, and those changes and refinements are a reflection of scientific progress.

Another view expressed by some is that, during scientific revolutions, whenever paradigm shifts occur, the choice between the competing paradigms is arbitrary. They look at the history of paradigm shifts and claim that, at the time when the choice of one paradigm over another was made, there was inadequate evidence to make the decision on more than an arbitrary basis. Of course, one can question the ability of those making this claim to be able to pick *the* point in time when the shift actually occurred. Not all scientists make the shift at the same moment.

More telling, however, as a refutation of such a view is the fact that the choice between a pair of scientific paradigms is far from arbitrary in the long run. Instead, a newly accepted paradigm is accepted because it explains more data than the old one. That is the main reason that scientists shift paradigms. Einstein's theories encompass Newton's—they are able to encompass *all* the observations of Newtonian mechanics. In addition, they explain and account for the anomalies that Newton's equations could not account for. Science does progress—it is able to account for more and more data. New predictions are made, which differ from those of the old paradigm, the predictions are tested, and the new paradigm holds sway only as long as those new predictions are not refuted by experiments.

It is obviously possible to pick a point of time during the competition between two or more paradigms when the evidence favoring any one is equivocal, and I suspect that the naysayers are doing just that. One must go on to see what happens after that moment—it is not as if no more experiments will be done. Most scientists may have a bias in favor of the new paradigm at the time a paradigm shift is occurring. They may even have decided, for themselves, that the new one is superior just because the most recent evidence appears to support it. However, at such a point, nothing prevents further evidence from turning things around. As with individual theories, if a mistake is made, and a weaker paradigm is accepted by a group of scientists at a given moment, further studies can cause it to be replaced or refined.

To demonstrate conclusively the superiority of one paradigm over another might take decades, or more, of further experiments. In the meantime, while the new paradigm "rules" for a majority of scientists in the field, the older paradigm might continue to hold sway among a minority of scientists. Even when there is no such minority, new evidence can create new views or even trigger the development of modified versions of old ones.

Consider the history of views about the nature of light. Is light a wave or a particle? The view among scientists shifted back and forth more than once over hundreds of years. We now think we understand why it was such a difficult issue. We had defined wave and particle as mutually exclusive options, whereas light appears to have properties that are both wavelike and particle-like, as discussed earlier.

Progress in science need not be straight-line progress—there are dead ends, and there can be runs in the wrong direction—but the history of science seems to be pointing in the direction of a more complete and accurate view of the world. We also are constantly testing the accuracy of our world views as we develop technologies that depend upon scientific theories. Furthermore, we can see a complementarity in the linkages and supports between scientific fields that have developed relatively independently, and this further strengthens the theories.

For instance, the findings in molecular biology, and the tools developed by molecular biology are being used today to confirm and extend evolutionary biology. We look at differences in the genes among different species and confirm the evolutionary distances between such species. The time from the last common ancestor can be estimated by the number of changes in the genome that have taken place. Another example of confirmation and correlation between scientific fields is seen in the estimates of the age of the earth and solar system that come from geology and from astronomy.

Further opportunity for confirmation and correlation has come from the development of new scientific fields that form bridges between existing fields, such as biochemistry, which bridges the sciences of chemistry and biology. We also have observations from physics and chemistry that are used as a foundation, and so tested in new ways, by molecular and cellular biology. In a similar way, there are multiple fields within biology, such as neurobiology and neurology, that are serving as a growing foundation for the development of a science of psychology.

It is very satisfying and reassuring to see the linkages, interconnections, and confirmations of existing theories and paradigms among different sciences. It leaves me with little doubt about progress in science. There is a growing understanding of the nature of the universe and an ever-larger data base that can be explained or understood by the theories and models of science, even when those theories and models undergo scientific revolutions. Kuhn, in his *Structure of Scientific Revolutions*, acknowledged that there is progress in science in the growth of data that can be understood and in the number of problems solved.

Revolution 3. Evolution and Natural Selection

This third revolution started with Darwin, but its consequences still reverberate 150 years later. We have confronted some of these issues as we considered *The French Lieutenant's Woman*. We can see our own society continuing to adjust to this revolution in the reactions of creationists and some other religious fundamentalists who attempt to control science content about evolution in public school systems. More recently there also have been attempts to claim the necessity for a designer for life. That is, some have argued that all of the complexity we see in living organisms could not have arisen without some kind of designer at the helm. Such arguments would appear to be made by those who have not understood, or have not been willing to accept, the role that natural selection plays in the evolution of complex systems.

For some it clearly has been an adjustment to accept that the earth we live on is four to five billion years old and that we are evolutionary "cousins" to other apes rather than being a special creation. Recently, even as important a religious leader as Pope John Paul II acknowledged that there is strong evidence in favor of the theory of evolution.[1] However, he also expressed concern about any attempt to develop a science of what he called the soul. I will address that concern later in this chapter.

Revolution 4. The Overthrow of Vitalism:
Life Obeys the Laws of Physics and Chemistry

I would like to make the case that biology has produced another revolution—a fourth revolution, whose finishing touches continue to be applied today. This revolution complements and extends the third revolution. In society, we still are confronting and discussing the implications of this revolution, which range from our views about hierarchies among nonliving and living organisms to the potential for genetic manipulation of humans and the possible impact that will have on human nature.

A simple view of this fourth revolution can be had by contrasting two perspectives on the nature of life. The classical view, as we have discussed, was vitalism, which held that living organisms were governed by special forces or spirits. There was seemingly such a great gap between living and nonliving—between animals and rocks—that life was thought to require something very different and special when compared with inanimate objects. The contrasting view holds that our knowledge of inanimate objects, and of the forces that govern them, forms an adequate foundation for understanding life.

Beginning in the 1800s and extending throughout the 1900s, we have come to understand almost all of the special properties of living organisms without the need for any new forces or vital spirits. Thus, the elimination of a need for vitalism has been a slow process of testing and understanding. The vitalistic expectation of the need for something new and different to explain life was found to be unnecessary. Physics and chemistry alone appear to offer a complete foundation for understanding events occurring in living organisms. Today, we have a good understanding of growth, metabolism, reproduction, heredity, movement, and sensation. Reductionistic approaches have been crucial in giving us our understanding of life and its mechanisms.

In a sense this fourth revolution continues the revolution of Darwin, but it actually started before Darwin and was only firmly established after his revolution was largely complete. As late as the end of the nineteenth century, a scientist as prominent as Louis Pasteur still held strongly to vitalistic views, long after Darwinian views on evolution had been accepted by most scientists.

The overthrow of vitalism probably is not as important a *scientific* revolution as the three discussed above, but I wish to identify it as a critical revolution in our thinking about the nature of life, and human life in particular. A little background about some formerly common cultural views might help clarify why this is a revolution in terms of its impact on human thought about human nature.

There is a rather influential, traditional hierarchical view of things: nonliving matter———> plant———> animal———> human—> divine, or spiritual.[2] This traditional view held that each higher level in this sequence adds some mysterious quality missing from those below. At each higher level new properties emerge that are incapable of being encompassed by the laws operating at the lower levels. While this view was shaken by evolutionary thinking, it took the shock of the demise of vitalism to fully dislodge it as a central part of a model of human nature.

Evidence from evolutionary biology dealt aspects of the hierarchical view an early blow. The proposed step from plant to animal is not consistent with the way evolution occurred. Both plants and animals evolved from earlier ancestors; animals emerged in the sea, where, to this day, single-celled microscopic animals consume other microscopic organisms. Animals only emerged onto land after plants were there, but animals had been evolving in water long before the land invasion. Today's animals and plants have common ancestors, if one goes back hundreds of millions of years. But neither can be said to have evolved from the other.

Also, with our understanding of photosynthesis, which occurs in plants and not in animals, it is not so obvious that animals belong above plants in a hierarchy due to special abilities. Thus, plants have special properties that animals do not have, as well as the other way around. It might be nice if humans were able just to take in carbon dioxide and a few minerals and bask in sunlight to generate food, rather than eating other living organisms to get it. Trying to arrange such different and varied life forms as plants and animals into an absolute hierarchy is not easy, and any attempted hierarchy may merely reflect the bias of the arranger.

A possible, scientific alternative to the hierarchical approach is to examine, and attempt to understand, the novel properties that emerge as complexity builds. Complexity does tend to be greater in living organisms than in inanimate objects, and some life forms, such as eukaryotic cells (those with internal membrane systems), are more complex than others, such as bacteria. Multicellular organisms also are more complex than single-celled organisms.

Even with all of the evolving complexity of life, scientists have found that the novel properties that appear with increasing complexity seem to result only from the interactions among components, and not from new, vitalistic forces. Molecules have properties that individual atoms lack, but it is the structural arrangement of the atoms in the molecules that yields the new properties and accounts for them. Groups of cells and tissues also have new properties over those of individual cells, as a result of the structures they form—consider the way that heart muscle cells combine to form the cavities that allow blood to be pumped. A simple, nonbiological example of what is meant by structural arrangement's yielding new properties, can be seen in the ability of a hammer to drive nails. That ability is not well represented in either the head or handle alone. It is the proper structural combination of the two that allows for a well-functioning hammer.

In a similar manner, the special properties of living organisms appear to arise from the organization of matter, not from mysterious new forces that go beyond physics and chemistry. Consider the special properties of water that make it so useful and necessary for life as we know it. For a small molecule, water has an exceptionally high boiling point and high heat content (it takes a lot of heat to change its temperature). These properties keep us from boiling away at room temperature and help us to maintain a more constant body temperature. Water also exhibits high surface tension and capillary action.

None of these properties are evident in the individual atoms of oxygen or hydrogen that make up water. So, how do the new properties arise? Can we understand these seemingly emergent properties of water on the basis of the physics and chemistry of hydrogen and oxygen? Yes. Oxygen tends to cling quite tightly to the (negatively charged) electrons that it shares with other atoms when it forms molecules. As a result, in a water molecule, the oxygen atom has a small negative charge, whereas each of the two hydrogen atoms linked to the oxygen are slightly positive. This allows neighboring water molecules to form linkages with each other. The (slightly) negatively charged oxygen in one water molecule attracts the (slightly) positively charged hydrogen atoms in nearby water molecules. Opposites attract. We call the linkages *hydrogen bonds*. These weak bonds *between* H_2O molecules are what give water its special properties. Because of hydrogen bonds, it takes more energy to break apart the water molecules, so water boils at a higher temperature, exhibits surface tension, and so on.

This use of the properties of parts of a system (in this case the oxygen atoms' attraction for shared electrons in chemical bonds) to understand the seemingly emergent properties at another level (the properties of water molecules) is an example of a reductionist explanation. It is not the only kind of explanation in science, but it has become an important source of understanding.

What is true of the properties of molecules compared with atoms is thought to be true on up the line, as macromolecules are built from molecules, as organelles are built, as cells are put together, and so on. The old hierarchy and vitalism appear to be wrong in expecting new, emergent properties unexplainable on the basis of the properties of the component parts and the structures they form.

During the past 150 years, as vitalism was found to be unnecessary, there were some high points that often are pointed to. For instance, it was once thought that organic matter, or molecules, could only be made by living organisms. The chemical synthesis of urea in the mid-1800s marked the beginning of the end for that view, and was the first real challenge to vitalism. With the development of such fields as organic chemistry and biochemistry, the need for vitalistic views became less necessary.

The quiet revolution has continued during this century as molecular and cellular biology have been able to account for more and more of the seemingly emergent properties of life on the basis of what is happening at a molecular and cellular level. We began to understand reproduction by looking at chromosome movements during meiosis. We went on to understand the molecular basis of heredity and genetic variations in terms of DNA and its sequence of base pairs. Animal movement due to muscles is understood in terms of sliding filaments made of actin and myosin proteins. We also now can map metabolism on wall charts containing the pathways of enzyme-catalyzed chemical reactions. In brief, we have rather good scientific explanations for what is happening with almost all of the special properties of life today. The last of the details are not in yet, nor will they be any time soon. But detailed outlines are unmistakable, and the answers so far all point to an ability to understand the special properties of living organisms on the basis of the physics and chemistry of molecules, their interactions, and the structures they form. There is no evidence of a need for anything more. There are no obvious "gaps" in our understanding of most of the properties by which life is defined.

Revolution 5. Mind from Matter: the Neuroscience Revolution

The success of the reduction of most of the properties of life to chemistry and physics sets up the possibility of what I will refer to as a fifth revolution—a revolution in our understanding of the relationship between mind and brain. You could consider this a continuation of the fourth revolution, but the consequences go deeper.

We have made great progress in understanding the physical and chemical basis of most aspects of living organisms, but what about mind? I take mind to include thoughts, perceptions, memory, feelings, emotions, and consciousness. Can we explain how mind works—how we think and what consciousness is—by looking at how the brain works? Is mind explainable in terms of physics and chemistry, or will this last bastion for vitalism present an insurmountable barrier for science? The question is important because the answer will influence not only ideas about the nature of free will, as discussed earlier, but also issues about whether the "self" survives death and other fundamental aspects of human nature. I am not the first to point out this potential development from the neurosciences.[3]

Let me begin with a brief history of views on relationship between mind and brain. The classical view was described well by the French philosopher Descartes, and has come to be known as interactionist dualism. This view presumes that mind is not a part of the material universe. There is a second "world" of mind, distinct from the physical world. Mind influences our behavior, and our sensory inputs influence mind. But mental phenomena are in a separate world. It has always been difficult to imagine how matter could give rise to experience, and this model of the relationship certainly is a comfortable one from this perspective. Interactionist dualism also can be comforting to those who believe in a life after death. Since our minds, our selves, or, loosely speaking, our souls, are not a part of the physical world, there would be no reason to assume that they must die with our physical bodies and brains. However, there is a serious implication for physical law that this view carries with it. Since nonphysical, mental influences on the physical world exist, there must be violations of physical laws whenever such influences occur. Whenever we "will" our body to do something, like speak what's on our mind, it appears that such an influence must disrupt the normal, lawful flow of cause and effect in the physical world. The disturbance of brain by mind would violate physical laws (Wilson 1976, 1999).

In the late 1800s T. H. Huxley, probably in response to such concerns about the violation of physical laws, championed another form of dualism that avoided the problem.[4] Huxley proposed that the traffic between the physical and mental worlds was all one-way. The brain was able to influence the mind, but not the other way around. The mind became an epiphenomenon, arising out of brain activity but unable to influence it. This view avoided the problem of physical law violation, but at an expense. Huxley's so-called epiphenomenalism left mind unable to act. Even our feeling of conscious decision making is only an illusion under epiphenomenalism.

There seems to be an obvious refutation of epiphenomenalism.[5] How can our voices (in the physical world) speak about our conscious experiences (in our minds) if our minds have no way of communicating with the physical world? What a grand coincidence, under epiphenomenalism, that we are able to speak in detail about the existence and nature of our conscious experiences since they have no way of acting on our voice.

Another argument against epiphenomenalism comes from evolutionary considerations. There would be no reason for painful things to hurt if pain played no role in the physical world. Natural selection during evolution would have been unable to influence whether experiencing pain feels bad and pleasure good were epiphenomenalism correct.[6]

So, both interactionist dualism and epiphenomenalism present real scientific difficulties as models for the relationship between mind and brain. More recently, monistic views have become more popular, among both scientists and philosophers, as models for how mind and brain relate to each other. Monistic theory views mind and consciousness as part of the functioning of brain: mental processes are brain processes. The "self" of each of us—our experiences, feelings, and so on—is a part of the physical world. There are a number of differing monistic views, the details of which need not concern us. Monistic views—mind is brain—have become the dominant model the neurosciences are working with. That is not the same as saying we have demonstrated that this model is the correct one, but it seems to be working as scientists begin to identify and define the neural correlates of conscious experiences.[7]

If some form of monism is correct, and there can be a science of mind and consciousness, then there will be a revolution in the thinking of at least some of us. I will call this fifth revolution the *neuroscience revolution*, since it is from the study of neuroscience—from such fields as neurobiology, neurology, and psychology—that this change in our view of the nature of mind and its relation to the physical world may come.

Some Data

Is it really possible that a brain can have experiences, be conscious, be a part of the universe able to observe and think about itself? Is the brain capable of being that complicated? I will try give a brief overview of a small part of the data that we have. There is no attempt here to be complete; if there is not enough evidence here to convince you that this is at least a possibility, try reading some of the references that I have listed.

Let me begin with a comparison between our brain and a typical desktop computer. In terms of the speed of units, the computer seems to have us beat easily. For a thousand dollars or so you can go buy a computer with a central processing unit that will carry out several hundred million calculations each second; that is, computer speeds have reached 800 MHz or more today. The units in our brains, our neurons, can fire electrical impulses, maximally, at about 500 times a second—about a million times slower. Then how are we any match for the computer? Answer: our brains do parallel processing. We are not limited to serial processing as the typical desktop computer is. We do not have to run all of our information processing through just one central processing unit. Instead, our neurons work in parallel, simultaneously carrying out processing of different parts of whatever task we are focused on at the moment. Now, computers have been built to do parallel processing, but the programming is difficult and our current desktop and laptop units don't do this.

How many such units does the brain have? The number is at least 100 billion, and some estimates range to several hundreds of billions, although making anything more than a rough estimate is difficult. To give you an idea how many that is, consider a television screen. The image on a television screen consists of small dots, or pixels, in arrays of hundreds by hundreds across the screen. The roughly 200,000 pixels on a TV screen can emit different colors at different degrees of intensity, and it is the sum of all of these little dots that produces the TV image. To give an idea of the enormous number of nerve cells in a single human brain, Paul Churchland has suggested the following comparison.[8] Consider one of the twin towers of the World Trade Center in New York City. Imagine that we cover the entire outside of the tower with television screens— just the picture tubes, positioned close together. That would require about 500,000 TV screens, each with almost 200,000 pixels. Now consider the number of pixels on all of those TV screens. They would be about as numerous as the lower estimates of the number of neurons in a single human brain!

Each of the neurons in our brain forms, typically, hundreds or thousands of connections with other neurons. There is an amazing complexity possible in the wiring patterns between the neurons. Those patterns are modifiable during development and learning, giving us a unique substrate for the complexity of our experiences. The possible momentary patterns of excitation across the brain are nearly limitless.

How far along are we in understanding mind? Neuroscience has developed preliminary models of some mental abilities, such as memory formation. Models of how consciousness works do not yet exist, but the search for the neural correlates of consciousness has begun. The search has been helped by some of the new, noninvasive techniques of viewing brain activity in humans, such as PET (positron emission tomography) and functional MRI (magnetic resonance imaging). We can follow the flow of analysis in different areas of the brain as we think, analyze, and see (McCrone, 1999).

We now know that such events as visual perception—seeing something—may involve a binding between events that are occurring in different parts of the brain. It is thought that simultaneous electrical impulses, repeating in unison about 40 times a second, in the neurons from different areas may play a role in linking things together. There also have been interesting insights into the nature of consciousness coming from studies of states of consciousness in sleep and dreaming, in comparison to our waking states.

Another example of our ability to begin to dissect consciousness comes from the experiments of R. W. Sperry and others.[9] In treating some individuals for epilepsy, the largest fiber tract in the brain, which links the left and right sides of the upper brain (the left and right cerebral hemispheres) has been cut. The resulting "split-brain" patients appear to have two conscious centers. The left side of the brain, which in most of us controls speaking, controls the right side of the body. The left side of the body is controlled by the right hemisphere, which is better at spatial tasks—reading maps, solving picture puzzles, and so on. The split-brain surgery appears to split our stream of consciousness into two somewhat independent flows after the surgery. For example, one such patient with more split language function was asked what kind of a job he wanted to have, the right hand (left hemisphere) wrote "draftsman," while the left hand (right hemisphere) wrote "race car driver."

How does the brain generate consciousness, feelings, perceptions, thoughts? How does the personal side of self get generated? How do we actually see red, feel pain, and so on? We don't yet know. We are starting to get some understanding of the neural correlates of these processes—to identify where in the brain such processes seem to be occurring, but the hard problem has not been solved. Maybe it won't be, and my argument will collapse. But let's assume that insights will come, and that I am right about this neuroscience revolution. What then?

What is the proper role of the humanities if we can expect to understand more of what humans are all about by studying neuroscience? What happens to the views of some modern religions, especially those that propose there is an afterlife or that there is reincarnation? We know what happens to the brain after death. It ceases to function in an irreversible way; it decays. Its capabilities are lost, its circuits destroyed. The recent findings of neuroscience, which support a monistic view of mind/ brain, suggest that as the brain decays, so does the mind. If the brain is essential for mind, how can mind or self remain? I do not intend this argument to diminish religion. Even if there is no afterlife, there are very important roles for religion in society.

Another indication of the revolution in the view of human nature involved in the idea of a physical basis for mind comes from the address that Pope John Paul II gave to the Pontifical Academy of Sciences on October 22, 1996. In it he discusses one of the earlier revolutions we discussed—evolution and natural selection. He states that "new knowledge has led to the recognition of the theory of evolution as more than a hypothesis. It is indeed remarkable that this theory has been progressively accepted by researchers, following a series of discoveries in various fields of knowledge. The convergence, neither sought nor fabricated, of the results of work that was conducted independently is in itself a significant argument in favor of this theory."[10]

But the pope goes on to say: "If the human body take its origin from pre-existent living matter, the spiritual soul is immediately created by God. Consequently, theories of evolution which, in accordance with the philosophies inspiring them, consider the spirit as emerging from the forces of living matter or as a mere epiphenomenon of this matter, are incompatible with the truth about man. Nor are they able to ground the dignity of the person." Notice here that "emerging from the forces of living matter" would appear to refer to materialist, monistic views of the mind-brain relationship—the very views of the fifth revolution. His statement "as a mere epiphenomenon of this matter" would be the T. H. Huxley view of mind. Assuming that what the pope refers to as the spirit or soul includes what I am referring to as mind and self, what we are left with as the only acceptable view of the relationship between mind and brain, according to the pope, is something like interactionist dualism. That view appears to be in direct conflict with the direction that science is heading, since few scientists continue to accept interactionist dualism as a viable view because of its apparent requirement for violation of physical laws. This is a problem not just for the pope and Catholicism but for many religions. There is nothing we now know that can "prove" science right or the pope right. For instance, he might have no problem with the need for violations of physical laws; after all, isn't that what miracles are all about?

From this one example, there should be little doubt about the major impact that a neuroscience revolution would have on our ideas about human nature. Consider also our earlier account of the idea of free will— our presumed ability to make totally free choices. As we discussed, if mind is brain activity, and nothing more, then our decisions appear to be locked by our genetics and experiences, and not to be free of physical causes. There would also be implications of this monistic view of mind and brain for our art and literature, and even for our legal system. Our sense of what we are will be changed profoundly if this neuroscience revolution is completed.

Bowen

I have little to quarrel with in David's narrative of the revolutions in science, particularly since they include as creative and imaginative a set of examples as any the human mind can construct—at least until the present mechanisms of the brain can be fully understood, controlled, and diddled with by present and future scientists. But I do not wish by such flippancy to demean the "revolutions." I cannot dismiss any of the revolutionary ideas, either those recently accepted, such as time duration relative to speed, or those now lying seemingly within the scientific grasp, such as the mechanistic character of human thinking. They may all be true or eventually come true.

David's narrative gains credence from its very structure, beginning with the historical past of the first revolution and then jumping ahead to the modern science of the last hundred years, and proceeding into the science of the future. The story is vintage cause-and-effect reasoning applied to the process of scientifically verifiable physical/chemical cause and effect itself. Rather than expressing horror at what the outcome might be for the dignity of the human race sometime in the technological future under the mechanistic manipulations of scientists and technologists, I will leave that to Huxley in the next chapter, and instead offer a couple of examples of how in their own way literary theory and fiction coincide with David's remarks on revolutions.

David's description of the new paradigms that arise from new complexities of a higher order was exemplified by the discovery that the molecular nature of water has different properties from its constituent atoms of hydrogen and oxygen. The same kind of reasoning applies to literary postcolonial theory, where the result of colonial occupation (or immigration for that matter) produces a hybrid of colonizer and colonized with properties different from either of its two main constituents. Homi Bhabha offers a metaphor of such hybridity in "Signs Taken for Wonders," in which he relates the story of five hundred Indian people sitting under trees reading the English Bible, the first book most of them have ever read. They believe the words of this wondrous book/sign and their remarks to an early Indian catechist who discovered them in the act indicated that they had absorbed the teachings of the book by quoting them as if they were in fact the word of God. But when asked why they were not baptized into the Christian faith, they claimed that they were unable to take the sacrament because "the Europeans eat cow's flesh, and this will never do for us" (30–31). Neither Hindu nor Christian, the hybrid Anglo-Indian is a unique entity, as resistant to complete assimilation as water is to boil at low temperatures.

The second example comes from a literary representation of David's idea of the materiality of man—including his perception, thought processes, and spirituality—as all ultimately subject, like any other matter in the universe, to the scientific laws of matter and energy. The passage comes from Heller's novel of human survival in an absurd world, *Catch 22*. Yossarian, the protagonist, is haunted throughout this tale of wartime stupidity and power-mongering by his memory of a comrade, Snowden, dying in his arms. He initially fails to recognize the severity of Snowden's wounds, and wraps him in a parachute which eventually parts, revealing in Snowden's disembodied entrails the "message" of the book:

> Here was God's plenty, all right, he thought bitterly as he stared—liver, lungs, kidneys, ribs, stomach and bits of stewed tomatoes Snowden had eaten that day for lunch. . . . [Yossarian] felt goose pimples clacking all over him as he gazed down despondently at the grim secret Snowden had spilled all over the messy floor. It was easy to read the message in his entrails. Man was matter, that was Snowden's secret. Drop him out a window and he'll fall. Set fire to him and he'll burn. Bury him and he'll rot like other kinds of garbage. The spirit gone, man is garbage. That was Snowden's secret. Ripeness was all.[11]

Yossarian's thoughts are the grim, existential results of facing what may be the hoped-for truth about the materiality of human beings and their meaning as life forms. Snowden's very name suggests Mount Snowdon, Wordsworth's ultimate embodiment of transcendent truth, but his final (dying) appearance in Heller's novel consists of a set of glutinous entrails on the floor of a plane. Is life pointless to the materialist, since it has no meaning in terms of some higher spiritual entity? Hardly. The very metaphor itself will inform Yossarian's clinging to life, making precious the very biological mechanism that gives it meaning. Far from degrading life's essence, it gives vitality to our efforts to exist on a scientific, empirical level of understanding. Being made of the same material as the rest of the universe, powered by the same diminishing sources of usable energy, and living always in the process of dying, does not necessarily diminish us, any more than our newfound knowledge of our own materiality. The metaphor of our vulnerable condition need not destroy anything but a false hubris concerning our supposed special relation to mystical ineffability. It might seem to degrade our species to learn that we were made in God's mechanical image, but there may also be some comfort in understanding who we are, and some satisfaction in discovering this disillusioning insight in a good novel as well as a biology lab.

Cycles and History/Yeats

Bowen

Our sense of history comes from several sources: our previous personal experiences, our genetically inherited knowledge or instinct, and the knowledge provided by artifacts passed on pictorially or verbally in some modified form. Together they constitute a continuum of cultural mythology, or, in more familiar terms, our present heritage. With the advent of literacy came written records providing more permanent accounts of events as perceived by writers over time and corroborated by other sources and documents.

Among the many lessons history provides are clues to our dealing with the present, and sometimes even a clue as to what might occur in the future. This idea of the past informing the future has permeated Western literature at least from Homeric times, when Odysseus visits Hades to discuss the past with the deceased, to gain advice as to how to conduct himself in the present, and, perhaps most important, to get an idea of what the future has in store for him. Aeneas makes a similar visit in the Roman epic, and Dante in his religious epic. What they (and the readers) learn from those who have preceded them is one of the principal causes of their successes and failures, and for the readers of these works, those conditions of previous civilizations and cultures that might either be accepted or rejected by their successive generations.

More recently, science, having now constructed an increasingly ample historical basis for comparison, has sought probability patterns for the recurrence of a variety of natural phenomena such as volcanic eruptions, tornados, and—important where I live—hurricanes. We know in these instances that the uncertainty of accurately predicting any single recurrence of a natural phenomenon at any given time is overwhelming, but the collection of historical data and its analysis have dramatically increased our prognostications and saved a lot of lives over the past few years.

Understanding the causes of historical events that humans precipitate is every bit as complex, in part because we think we can do something about them, that we can influence the history of the human race in ways that, until very recently, we could not do with nature. When it became apparent during the past few years that human activities were influencing natural calamities by creating such conditions as overpopulation, global warming, deforestation, acid rain, species elimination, and so on, the old faith in the immutable prevailing forces of nature still comforted the rationalizations of short-run profiteers, but the calamitous prognostications of environmental scientists are even now being realized.

Collectively we feel that by understanding what has happened in the past we can imitate those activities and decisions we think were good, and avoid the activities that brought about calamities. We try to make sense of history by applying the quasi-scientific axiom of cause and effect. While that doesn't work very well with unique new situations—like the long-range effects of the introduction of weapons of mass destruction, the development of instant communication, genetic manipulation, and a host of other scientific innovations whose ultimate effects are uncertain—we continue, full of hope (and occasionally, despair), to throw the dice of discovery, hoping to win.

Although it is possible that the cause-and-effect lessons of history may provide some hint of what we are about in the present, the more closely we examine the causes of such manmade events as the twentieth-century wars in which our country has engaged, the less we are likely to agree. One popular adage of partisan observers is to leave the legacies of presidents and other world leaders (except for Ronald Reagan) "to the historians to decide" who was at fault, who were heroes, whose faces should be carved on Mount Rushmore or have airports named after them. Military leaders have traditionally read history as cause and effect—what Hannibal or Caesar did in such and such a battle—as if there were not a host of contingent factors involved in each engagement.

The prime example of what can go wrong by applying simple cause-and-effect reasoning to war was Vietnam. No clear-cut bombing of the United States fleet prompted the Vietnam conflict, as it did at Pearl Harbor. Rather, a government-manufactured report (or lie) regarding aggressive action against American ships in the Gulf of Tonkin gave Congress an excuse to enact the Gulf of Tonkin Resolution, committing our country to all-out war. During and after the war, conservatives blamed the antiwar people for causing and prolonging the conflict, which, they claimed, would have ended in victory had we nuked the whole country off the map, while the antiwar people blamed the government for pursuing an illegal and immoral war in the first place. And that difference is nothing to the World War II cause-and-effect sequence that dismisses the Holocaust and deifies Hitler, a popular scenario among the good folks of the Aryan Nation.

The point is that there is no single meaning in the various narratives. We selectively hear what we like and accept it as fact. Let me explore three historical scenarios, the single unified evolutionary continuum, the dialectic progression of cause and effect, and the allied idea that casts history into a series of recurring cycles. Those of us in literature who still attempt to link the succession of ideas through time as a means of bringing coherence to our ethnocentric concept of the Western world—excluding all the rest of the world's civilizations—see evolutionary progress in our own version of intellectual history. The concept of a single historical narrative developed in the introductions to period anthologies purporting to be surveys of Western Literature is a largely academic attempt to create an official history, a tradition in which we and our students might be comfortably assimilated, often bearing titles that include terms like "Literary Tradition" and "The American Heritage." The books in effect create a master historical narrative explaining and containing thousands of individual narratives, chosen to fit the pattern. Of late, anthologies of narratives counter to the diminishing literary hegemony feature the neglected or subjugated knowledges we discussed in the Foucault chapter. These counter-hegemonic narratives often fall into a master narrative pattern of oppression and discovery of counter-truths to oppose the Western heritage party line. All of these narratives have, like their predecessors, their own political spin.

Some of what I have already said about Judeo-Christian literature in the Origins chapter of this book can be seen in the following tongue-in-cheek, grossly oversimplified, and perhaps distorted, one-page sample of the master narrative of traditional Western Literature/Intellectual History courses generally offered over two semesters:

The old monotheistic testament established a code of conduct, discussed the nature of sin, and justified the ways of God to man. Its counterpart in humanism, or a belief in the ascendancy of mankind, was given voice by the Greeks, who cut hecatomb deals to propitiate any number of gods, even as men went their creative ways. Their coping experiences gave rise to independent heroes who faced their own inadequacies and sought secular answers in order to bring light to a civilization in which they and not the deities were the protagonists.

The Romans conquered and appropriated the Greek civilization, adding still another, more stringent, patriarchal authority that manifested itself in their engineering, codification of laws, and submission to the authority of the state.

The Christian religion appropriated the secular rule of the Romans, adopted the idea of a cumulative ecclesiastical history and began codifying their belief system as a hybridized monotheistic code. The uneasy alliance between church and state manifested itself during the middle or "dark" ages, in which human intelligence and perfectibility retreated into a dark corner of religious servitude.

The Renaissance, or Rebirth, brought with it a return to classical and humanistic thought, and Western literature became once more a celebration of man, while eventually the Age of Enlightenment pushed the limits of rational reasoning processes to new heights, culminating in the civil rebellion which began the Romantic period, with its unruly questioning of authority, and its emphasis on emotion over reason.

The later nineteenth century brought a return to conformity, social, religious, and political, epitomized by the era named after the dowager queen of England, Victoria. All this gave way to the underlying libido of her son Edward and his adherents, and the hypocrisy that led to the two world wars of the twentieth century, each followed by a period of despair, called alternatively, "The Lost Generation," "existentialism," the "Beat Generation."

This brief chronicle, immensely flawed by omission, misinformation, and idiosyncratic interpretation, represents something like the skeleton of intellectual history on which we have hung all the literary baubles we could get into our courses. I doubt that one Western Literature teacher in ten would subscribe to its sweeping generalities of cause and effect, any more than I would. What I have chosen to represent here is merely an arbitrary attempt to give chronological coherence to the thousands upon thousands of literary works we might have taught if we had read them and had the time for them in the syllabus. And counter-culture antihegemonists are unearthing every year more buried or subjugated manuscripts whose potential addition to the syllabus dwarfs the former canon.

The second construct of history I would like to mention is dialectical: the presence of opposite voices/opinions/belief systems, which together set up a dialogical discussion of countervailing ideas.[1] Championed by Hegel, it features a thesis (idea) and antithesis (counter-idea), ultimately resolving themselves into a synthesis or hybrid blend. Marx applied the idea to economics and class differentiation, setting up the thesis as the prevailing rulers and the antithesis as the lowest class. Thus the progression of history was represented by a dialectic: feudal system rulers (thesis) versus the serfs (antithesis), with an evolving synthesis and new thesis in the middle class (bourgeoisie). These were opposed by a new antithesis, the proletariat (workers), until the final synthesis, a classless society, emerges.

A corresponding linguistic application of dialectic by Mikhail Bakhtin involves a dialogy or dichotomistic opposition of different voices and opinions producing a work of art. Another philosophical application of the dialectic process to history in general involves the conservative hegemony who favor the status quo versus its antithesis in the revolutionaries who demand change. Conservatives of whatever persuasion are usually for the existing government/economic system, and the liberals/revolutionaries (whigs, radicals, or whatever) for change, sometimes mildly democratic, at other times radical and even bloody. This dialectical division has historically occasioned so much conflict that it becomes another paradigm for history itself.

The third view of history I would like to mention is cyclical. Most of us would agree with Oswald Spengler and his followers that history as a whole runs in cycles, with analogous or similar events and constructions recurring over a long period of time. The notion has both a neat aesthetic appeal and an affinity with cycles in nature—solar, seasonal, tidal, life-to-death, reproduction, and so on—events that punctuate the passage of time.

I do not mean to imply that all historical cycles are stretched out over millennia. Mini-cycles occur frequently. For instance, in academia, about every eight years the faculties of colleges and universities get together and reinvent the core curriculum requirements. Every ten years or so we debate the efficacy of student evaluations of our teaching with all the same comments pro and con reiterated in connection with salaries, popularity, grade inflation, and all the usual suspects. The process is a lot like El Niño. There is no immediate rational reason we can understand for its repetition except that something like climatic changes occur with enough frequency that we can name the phenomenon without fully comprehending why it so regularly occurs.

Usually in the longer course of human events many of us think of history as punctuated by millennial cycles, as we have so recently become aware. The importance of the millennial idea itself has been intertwined with our Western concept of time. The present connotations of the term *millennium* as a watered down concept of the span of a thousand years did not always apply. The earlier Hebraic idea of the word embraced a thousand-year period, immediately preceded by an apocalyptic discord and the appearance of a messianic figure to cast out evil and to rule in peace and harmony over the righteous—separated from their evil counterparts—for a thousand years. After that the Messiah would bring back the evil for another encounter, when ultimately the permanently revanquished evil would give way to general celestial bliss. As early conceived, the notion, like other Jewish biblical symbology, was political in that it addressed the need for a safe, prosperous home in which the living might be gratified as well as the righteous deceased. It lent tradition and prophecy to the concept of a beatific hiatus for the Jewish people, continually dispossessed by history and their monotheistic creed.

The notion also proved an effective precedent for beleaguered first- and second-century Christians, who assigned the messianic role to Christ, and whose sectarian ascendancy in the religious hierarchy was slow and painful in coming. The more sects and the more nations who adapted it, the more variations were applied to the mythology. Most noteworthy was Augustine's idea that the millennium had already begun, and was on the ecclesiastical road toward ultimate salvation. As the centuries passed, the idea of making some kind of prophecy regarding the afterlife more immediately available to living people continued the political aspects of its Jewish forebears, and contemporary historical events, disastrous as well as benign, gave rise to a plethora of interpretations, revised and revitalized through time and place over the intervening centuries, as Catholicism became more politically complex, and later, when the Reformation opened the gates of interpretation to history, belief, and clerical power politics. Yet the idea of apocalypse persists right down to contemporary fundamentalist evangelism and its dire prophecies.

I have chosen some of the poetry of William Butler Yeats as a singularly creative narrative of the cycles of history. Yeats did not base his ideas on scientific observation so much as messages from the spirit world, in fact through a process of "automatic writing." His wife became the medium through whom Yeats claimed to have received information regarding the cyclical nature of the universe, information he compiled in a book called *A Vision*. While David and other scientific mentalities might greet this information with some incredulity, it did provide a foundation for some of the most striking poems in the English language, themselves candidates for appreciation and marvel throughout the next millennium or two.

Yeats's cycles were approximately two thousand years in duration. He conceived of them as beginning with a sign, a metaphor, or an image of a significant event in time, civilization revolving around that image but getting further and further removed from it over the years, until chaos ensues. The idea leans heavily on Christian millennial apocalyptic imagery, but accompanies each cataclysmic conclusion to an era with the birth/inception of a new driving image to replace the former one. Yeats saw his own time (the early twentieth century) as ripe for such a new ending/beginning:

The Second Coming

Turning and turning in the widening gyre
The falcon cannot hear the falconer;
Things fall apart; the center cannot hold;
Mere anarchy is loosed upon the world,
The blood-dimmed tide is loosed, and everywhere
The ceremony of innocence is drowned;
The best lack all conviction, while the worst
Are full of passionate intensity.

Surely some revelation is at hand;
Surely the Second Coming is at hand.
The Second Coming! Hardly are those words out
When a vast image out of *Spiritus Mundi*
Troubles my sight: somewhere in sands of the desert

A shape with lion body and the head of a man,
A gaze blank and pitiless as the sun,
Is moving its slow thighs, while all about it
Reel shadows of the indignant desert birds.
The darkness drops again; but now I know
That twenty centuries of stony sleep
Were vexed to nightmare by a rocking cradle,
And what rough beast, its hour come round at last,
Slouches towards Bethlehem to be born?

Our current civilization is like the falcon circling ever farther away from its original inspiration, losing sight of the image on which it is based. Yeats's world (he wrote the poem in 1920) was reeling in the throes of post–World War I trauma and the Irish Rebellion of 1916 with its aftermath of continuing civil strife. The first stanza describes in very realistic terms what the conditions were like. The persona of the poem is looking for a revelation, conditioned by Christian millennial narrative to expect that one will appear in the chaos of the time. The vision he sees out of the collective memory of the past is not the new image for the future, but a four-thousand-year-old image of the Sphinx, the metaphor of the ancient civilization before the time of Christ. There is sensuality in the slow moving thighs and indifference in his blank, pitiless gaze. The indignant desert birds bring us back to the original falcon, now out of control.

The vision fades, but its message remains. The Egyptian image was two thousand years ago superseded, "vexed to nightmare," by the "rocking cradle" of Christ's birth. Now another two thousand years later, the persona wonders "what rough beast, its hour come round at last, / Slouches towards Bethlehem to be born?" The term, "rough beast" precedes the immortal word, "Slouches." The image is not exactly a traditionally beatific one. The glory of the poem resides in no small part on the prospective ignominy of the new image itself. With Christianity reduced to a rocking cradle, and Egyptian and Greek civilizations represented by an insentient beast, what new scruffy metaphor will we pay homage to in the next two thousand years?

While there are a number of Yeats's poems that deal with cyclical history, I want to leave you with the one that asks the big question about the cyclical process itself. "Leda and the Swan" provides a different image for the beginning of the Greek civilization. The inception of the classical era is the literal, albeit figurative, image/narrative of Zeus, metamorphosed into a swan, inseminating Leda, who bore Castor, Pollux, and Helen—the conquering of Troy and the beginning of Greek civilization—as its issue. The poem is one with which feminist critics early took issue because it seems to glorify rape. I am not going to defend Yeats's libidinous choice of subject matter or the graphic gratification it provided the poem's dirty old bird or Yeats himself, but I want to concentrate on the question asked in the last stanza.

Leda and the Swan

A sudden blow: the great wings beating still
Above the staggering girl, her thighs caressed
By the dark webs, her nape caught to his bill.
He holds her helpless breast upon his breast.

How can those terrified vague fingers push
The feathered glory from her loosening thighs?
And how can body, laid in that white rush,
But feel the strange heart beating where it lies?

A shudder in the loins engenders there
The broken wall, the burning roof and tower
And Agamemnon dead.
                          Being so caught up,
So mastered by the brute blood of the air,
Did she put on his knowledge with his power
Before the indifferent beak could let her drop?

If the knowledge of Zeus, a knowledge of antiquity preceding his ascendancy to the throne of Kronos, his predecessor in an even earlier cycle, were passed on through Leda to the Greeks, and again to their Christian successors in later epochs, then human knowledge might very well be as cumulative as we would like to think. The world may very well not be reduced to infantilism with each ensuing civilization, but increase in knowledge as well as power. Another way of putting the question is, "Did we—will we—ever learn anything?"

Wilson

In giving several different views of history, each inadequate to cover everything, Zack seems to recognize the futility of any simple attempt to capture the full complexity of human history, and the great risk of oversimplification. However, he seems to think there might be something to the idea of cycles in history, so I will add a disclaimer of sorts, just in case.

As Zack might say of my suggestion that we are close to the truth with some of our scientific theories, I now can say of any cyclic view of history that life is not that simple. One must be careful not to pick and choose a few facts or events that appear to fit a pattern, a spiraling upward like a falcon, while ignoring the many other facts or events of history that would contradict such an oversimplified view as that of a one- or two-thousand-year cycling. The idea of cycles is just too simplistic to account for the complexity of civilization and its history. I more often see the vulture than the falcon spiraling upward, reminding me of what happens to dead theories.

At the same time, Zack's hope that we do not return to infantilism with each of his cycles has an interesting counterpart if applied to Kuhn's paradigms. To the extent that one accepts Kuhn's view of science as a cycling of paradigms (probably also an oversimplification), one can ask whether, as each scientific revolution occurs, we return to "ground zero," and have to begin all over? Some who argue that science does not progress seem to be making such a claim. In our earlier chapter on scientific revolutions, I tried to indicate why that view is wrong, and how there is good evidence that there is progress in science even with the shifting of paradigms.

12

Ethics and Morality
in Science and Technology

Wilson

There are two aspects to ethics and morals that I will deal with in some detail and a third that I will discuss only briefly. One aspect concerns the special issues of ethics and morality in the practice of science. A second concerns the ethical and moral responsibility of scientists within the broader context of society. The first is an inside view of ethics within science; the second is the role of scientists in society. The second quickly will get us to a consideration of technology—the application of scientific knowledge.

A third issue concerns the possibility of a science of ethics—can we scientifically determine what is moral and ethical? I don't think so. Science can tell us what *is* but cannot tell us what *should be*. Science speaks to what is possible, not what is right or wrong. Science can provide practical knowledge about the world and about ourselves, and we may use this knowledge to guide us in moral and ethical decisions, but the decisions as to what is ethical and moral, what is right and wrong, do not seem to be ones that can be decided merely by doing an experiment.

I think that some sociobiologists have exaggerated what a science of ethics and morals can achieve. For instance, some question how altruistic behavior by one individual toward another who is not closely related (genetically) can have evolved. Where is the selective advantage in, say, risking one's life for another, unrelated individual? I suspect that looking for a full justification of such behavior in evolutionary theory is looking in the wrong place. While human mental abilities and the foundation of human behavior lie within our evolved brains, nothing demands that every human act contribute to survival or reproduction. Human language and thinking abilities may have developed because they gave selective advantage to humans, but once developed, such abilities need not always promote Darwinian fitness. One need only consider teenager suicides to realize that our behaviors are not always driven by Darwinian fitness.

We may find some biological explanations for why humans have developed certain moral positions, and we may find biological explanations for why some moral standards are difficult for some of us to maintain. But a full understanding of morals and ethics, and their continued development, will necessarily go beyond science alone.

Ethics within Science

Personal Ethics of Scientists

There are expected ethical aspects to the behavior of scientists: honesty about reporting the results of experiments is an obvious one. Nevertheless, not all scientists are honest. Scientists are like other groups of humans: there are all types. There are examples of scientific dishonesty, from Piltdown man, a skull of the "missing link" between humans and apes that turned out to be a fake, to a more recent fabrication of skin growth, where a Magic Marker was used by a scientist to make some of the fur on a white mouse appear to be the result of a black skin graft.

Does the existence of intentional dishonesty destroy science? It might do serious damage, and make the practice of science quite difficult, if it were widespread. From my personal experience, cases of intentional fraud are relatively rare. Most of the papers in the literature are honest attempts to present real data and analysis.

Most scientists realize the importance of honesty in experiments. Most are aware of the risk of even unintentional bias messing up their results, and most appear to be doing their best to be honest. In the end, what is most important is "getting it right," and anything that interferes with that will cause one's reputation to suffer. Even unintentional bias can do a lot of damage to an individual's reputation and to the acceptance of one's work; outright fraud and dishonesty will usually destroy one's reputation. A few scientists may get away with things for a while, but in science, those who cheat or who fabricate results usually are found out eventually. That is because, if a scientific result is at all important, others will attempt to repeat it. A scientist with a reputation for getting results that others cannot reproduce pretty soon finds that his or her papers are not read by other scientists, and grant requests go unfunded because of peer review—if the person can keep a job at all.

There is another unethical behavior in science that perhaps is not dealt with as harshly as it might be, and that is the stealing of the ideas of others. Twice in my career, other scientists appear to have taken ideas of mine, gone back to their labs, done the same or related experiments based on my ideas, and published results with no mention of the idea having been mine. This unethical behavior is probably much more common than fraud. This is especially worrisome because of the way science is funded. Research grants are necessary for so much of science, and to gain such funding, one is required to put one's plans and ideas down on paper. Usually these grant proposals are read by peers, people who are in the same field, capable of doing the same experiments. What is even worse, a dishonest reviewer can pan the proposal and thereby delay the person who had the idea from proceeding as quickly. It becomes very difficult for the wronged scientist to "prove" that the idea was hers alone. The other individual can argue that he got the idea on his own.

There are other more subtle issues related to ethics and morals within science that also can be important. As an example, consider the recent incident involving "cold fusion." Two groups of scientists independently submitted papers and held press conferences to announce that they had developed a system that produced energy from the fusion of atomic nuclei at room temperature. Many groups have been attempting to develop fusion energy for years. Fusing atomic nuclei is the kind of thing that occurs in the sun—at very high temperatures and pressures. Unlike others who have been working on the problem, these two groups each claimed that it had nuclear fusion occurring at room temperature in a relatively simple apparatus. They were wrong.

This, however, does not seem to be a simple case of fraud. The scientists had actually deceived themselves. They had not done all of the necessary experiments to demonstrate what they claimed. After their announcement, others tried to reproduce their findings, but failed. Their claim produced headlines around the world. If they were right, cheap, virtually pollution-free energy was soon to be available to everyone. What caused this fiasco? It appears that a mix of things were involved. One was the desire of the scientists to be first—the two groups viewed themselves as being in a race with each other. They had a side agreement that they would publish together. One broke the agreement and submitted a paper without telling the other. The other rushed to publish as well. News conferences were called by the scientists' home institutions to announce the great results.

Both research groups and their universities were expecting big dollars for grant support and royalties—fame and fortune. If cold fusion had really worked, the first to show and publish the result would reap big rewards—this was Nobel Prize stuff, the finding of the decade or of the century. Usually this would lead to great caution on the part of a scientist. One wants to be *very* sure before risking publishing. In the back of one's mind: what if I am wrong. No one will ever forget, I will be embarrassed, ruined.

In the case of cold fusion, greed and the rush to be first overcame caution. It was a marvelous feeding frenzy by the press—both at the time of the announcement and in its aftermath. At the time of the announcements, the two universities even joined in the fray of debating who was first with the "discovery."

As this example indicates, in recent times the doing of science sometimes has taken on unfortunate characteristics all too common in our general society. Not all scientists are willing to be so incautious as those involved in cold fusion, but I'm also sure that these are not the only ones.

Ethics in the Performance of Experiments

There are other ethical and moral issues that arise within science in the carrying out of experiments. Before the advent of human subject committees at universities, and an increased sense of responsibility toward the subjects of human experiments, there were excesses—moral lapses that are obvious by today's standards. Two examples come to mind.

In Tuskegee, Alabama, a group of poor Southern blacks with syphilis were part of a long-term study of the effects of the disease. During the extended experiment, new treatments for syphilis, based on antibiotics, were developed that could cure the disorder. The individuals who were being studied were not told about the cures and were continued in the study, getting sicker and sicker. I cannot imagine what was in the minds of those involved in such an inexcusable continuation of the experiment.

Recently, the United States government announced that experiments in the 1950s had exposed individuals, including children, to low levels of radiation without informed consent on the part of the individuals in the studies.

The frequency of such ethical blunders in experiments on humans has declined greatly in the last decade or two, as standards have been developed for informed consent by the subjects and as human-subjects committees, which review research plans before experiments are permitted to start, have been mandated. However, all of the issues have not been resolved and some are not so simple, as ethical dilemmas can still arise. For instance, how do you get informed consent from an unconscious individual? If you can't, how can you ever study certain disorders, and the best treatments for them, when the individuals with the disorder always arrive at the emergency room in an unconscious state?

Another issue arose recently with AIDS research being carried out on subjects in Africa. Since many Africans with AIDS cannot afford the medical treatment, which currently involves taking multiple drugs to suppress the virus, what was the proper treatment for the "control" group—the group not getting a new, experimental drug? Since these individuals normally went untreated anyway, could the control group be given nothing, or should they be given the "best" known treatment used in wealthier countries? If the latter, then the experimental group would be given that treatment *plus* the new treatment. The choice might appear to be clear, that the latter course should be followed, but consider this dilemma: what if the new treatment was most effective only if given alone, without the other drugs? The desire to test that possibility caused a recent study to be done in Africa, and the original design for the experiment was, indeed, to give nothing (beyond a placebo) to the controls and to give only the new drug to the experimental subjects. There was an international protest about the initial design of the experiment.

I think that there have been extremes in ethical demands by some. For instance, there are some "animal rights" activists who have declared that humans are no more important than other animals, and that we should do no experiments on animals. Were such a limitation to be imposed, advances in medical sciences would be greatly slowed, or in some cases stopped, by the inability to use animal subjects.

I once debated an "animal rights" activist from PETA, People for the Ethical Treatment of Animals, on a radio talk show. Early in the program, the talk show host asked, "If a pigeon and a human baby both are in the middle of the road, with a truck about to hit them, and you have time to rescue just one, which should it be, the pigeon or the baby?" When the person from PETA declared that it did not matter, since the two living beings were equally important, the talk show host and call-in audience got so outraged that I needed to say very little more for the rest of the show.

I believe that we have obligations and responsibilities toward other animals, but that most other animals do not have "rights" in the same sense that humans do. I think it is important to be caring about animals and to cause them as little pain and suffering as possible during experiments. There are no alternatives to using animals in some experiments, if we are to advance the science of medicine. There are times when one has no choice if one wishes to know something that might benefit humans and animals in the future.

Rights come with and from some of the special characteristics that make us human—the ability to think and reason, to be self-aware, to have a well-functioning mind. We seem to recognize this even in modern definitions of death. If the brain is dead, so is the person. We remove (still living) organs from such brain-dead individuals to use for organ transplants.

Along with rights come responsibilities. Most other animals would seem unable to deal with obligations or responsibilities. Their behaviors are not usually judged as being moral or immoral, as are human behaviors.

If there are animals that might have rights, they would include our close cousins, the chimpanzees and apes. These species have minds that may not be developed to our level, but may be close enough to warrant special consideration. We are still studying the minds of such other primates, and do not know all of their capabilities. Some may argue for a few other species having the special mental capabilities, but the list would be short.

A final comment arises from the recent arrival of letters containing razor blades at the laboratories of a number of scientists who study animals. Fortunately, the razor blades did not cut anyone, despite being positioned along the top edge of the envelope so as to cause cuts. Animal rights activists claimed responsibility. There seems to be something wrong here. How can someone claim not to wish harm to any animals, and then mail out razor blades? Have they forgotten that humans are animals, too? If they feel justified in doing such things to humans, would they be willing to do harm to carnivores, such as lions and ospreys, which kill other animals? If they were to argue that there is something special about humans, that they should "know better," then they have given up on the argument that all species are equal.

Ethics between Scientists and Community

Choice of Experiments

A much broader issue concerns the responsibilities of scientists toward society as a whole. Do scientists have a responsibility, perhaps above that of many others, to be involved in community, to contribute to debates and discussions concerning the development and use of technologies that might result from their basic research? Do scientists bear responsibility if others develop harmful technologies from their basic research? (Do they not share in the glory if good technologies are developed?)

I will consider one example from the 1940s. The development of the atomic bomb during World War II is a very telling moral tale. Let me set the stage. Hitler's Germany was gobbling up the nations of Europe. Hitler was a totalitarian monster, hell-bent on killing all Jews and ruling the world. The United States was at war with Hitler's Germany and Japan.

Scientists in our country, including some who had escaped from Nazi Germany just in time, knew that some of the scientists who stayed in Germany had knowledge of energy release from atomic nuclei. Were the Germans trying to build an atomic bomb? Dare we not proceed to do the same? The decision at the moment seemed an obvious one, and the United States began a massive effort to build atomic bombs. The best book I have read for a detailed history is *The Making of the Atomic Bomb*, by Richard Rhodes. The war in Europe was over by the time we had built atomic bombs, but the war with Japan was still going on. By that time, a number of the scientists who had worked on the bomb had misgivings about any use of the bomb. Some tried to block it, but by then the decision was in the hands of the politicians, not the scientists. Immediately after the war ended, a number of the scientists pushed to limit atomic weapons. Some recommended sharing the knowledge with the Soviet Union. Some opposed the further development of the hydrogen bomb. Some formed groups and conferences to discuss nuclear disarmament and propose ways of stopping the testing of atomic weapons as the cold war developed. Some felt that they now knew sin.

The development of the atomic bomb actually involved a group, mostly consisting of formerly basic scientists, called into action during wartime to work on applied technologies. That is not unusual in time of war. During World War II there were other scientists who worked on the development of radar, and others who worked on cracking some of the secret codes that the Germans were using.

Moral and ethical issues can be complex ones, and can look different from different perspectives. There is no exact science of morals, nor will there ever be, but I do believe that there are differences in the moral issues for basic science and for technology, and these we will discuss shortly. I also believe that we humans can, and often do, make progress in the development of moral and ethical views and standards.

There are other issues as well. When a scientist chooses a research area, should that person be guided by the possible social applications of the research? Should the choice be made on the basis of what society needs? In one sense this question has been answered for many scientists—the federal government has become the chief source of money for the experiments that many basic scientists do, and as such, the kinds of experiments being done are specified, at least in a general way, by the decisions of government, especially our elected members of Congress, on what research to fund. Some things get a lot of money—cancer research and AIDS research have been popular recently. Other kinds of research, which might aid people elsewhere in the world, like research on malaria or river blindness from black fly disease, receive much less in the way of funding from the government. This actually is a moral issue for us all— should we just emphasize research on disorders that we have here in the United States, or should our priorities for spending research dollars be influenced by how common certain diseases are worldwide? Occasionally, a subject for research has such high social significance, such as studies of homosexuals, or of race and IQ, that members of Congress will attempt to interfere with the research, and block its support.

These are issues for society as a whole to ponder. Why can we afford to spend money on high-tech approaches to heart transplants when we do not spend enough on prenatal and infant care in our society? How can some of us spend money to buy cigarettes, and some of our taxes be spent to subsidize tobacco farmers, when cigarettes have been demonstrated to produce death and disease in society?

Science versus Technology

I think it is important not to confuse science and technology, because moral and ethical arguments can be quite different, depending on which of these one is talking about. I consider science to be the pursuit of knowledge—knowledge about the world, the universe, as in the fields of physics, chemistry, biology, psychology, and sociology.

Scientific research can pursue knowledge for knowledge's sake, with less of an emphasis on a particular application, and I will refer to this kind of science as *basic research*. An example of basic science is the discovery of the structure of DNA by Watson and Crick.

There also is a more gray area, between science and technology, in which scientific research is more *applied*. An example of applied research would be research on AIDS patients, testing whether a new drug helped prevent the development of AIDS. It is very common for companies to be engaged in applied research targeted to a particular human problem, with the goal being the development of a new product or treatment.

Companies often do research and development related to new products—pharmaceuticals are tested on animals, on humans for risks and side effects, and then on humans for effectiveness in treating a particular disorder, as an example.

Moral and ethical issues arise especially in the applied-science, or technology, realm—the direct attempt to make new products for human use—new cars, houses, microwaves, more rapidly firing AK-47s, faster computer chips, more deadly biological weapons, better chainsaws for cutting down rain forests, more efficient processing of wood and pulp for production of paper for novels or journals, higher resolution television—the list is almost endless. When it comes to technology, there is little that does not carry with it moral and ethical implications. Usually such implications are complex ones, but sometimes they seem quite simple.

From where I stand, the decision of whether to work as a scientist or technician for a tobacco company seems a simple one. But how direct does the harm have to be before a moral issue arises? What about designing new cars? Should you be willing to design or work on the construction of sport-utility vehicles, pickups, and the like, when it has been shown that in collisions with cars they are more likely to kill someone in the car than would be the case if two cars collided? Do you, as a designer or builder of such vehicles, share a moral responsibility when someone driving your product kills another human? If you had not built them, would the person in the SUV perhaps have been driving a smaller car, and would the driver of the other car then have survived?

The consequences of a new technological product cannot always be foreseen. Sometimes the unforeseen consequences can be good ones, sometimes bad. I doubt that Henry Ford thought much about the risk of the greenhouse effect and global warming when he developed the production line that made cars more affordable, and hence more abundant.

The guy working on new glues apparently didn't have Post-It notes in mind until he developed a glue that didn't work very well—perfect for removable stick-ons, which at least some of us consider to be a useful product. Sometimes you do not know whether what you are doing will turn out to be good or bad until it does.

With many products, whether they are good or bad depends on how humans use them. In the end, we probably need to realize that it is the end user who makes the ultimate moral or ethical decisions, but it would be too easy to leave those involved in the invention, development, production, and marketing of new products blameless. I certainly don't hold tobacco company executives blameless—or tobacco farmers and the Congress of the United States for voting to subsidize them.

In comparing basic science with applied science and technology, I would suggest that we need as much basic knowledge as possible, but should apply that knowledge with caution to avoid doing more harm than good. We need basic knowledge so as better to understand and predict the results of the introduction of new technologies. However, with our limited knowledge base, some unintended consequences only become apparent afterward. That suggests we would be safer with a slower application of new technologies. Unfortunately, in our society, there seems to be a push to develop new technologies as quickly as possible—there is a big premium on being first, and the potential for big rewards. So long as this remains the case, we will have to live with the resulting consequences, and hope there are no great catastrophes.

I don't mean to suggest that I have given this topic its due, but hope that I have provided a preliminary sorting of some of the levels and issues relating science and technology to morals and ethics. We all make decisions every day, and are not always aware of all of the moral and ethical implications of our decisions. We choose a company or organization to work for; we choose a life style; we decide how far to commute to work by the location of our work and of home; and we often spend all too little time thinking of the many consequences our decisions have for others and for the world. Perhaps it is not so hard to imagine that engineers, technologists, and scientists often do the same.

I view science as a quest for understanding, not power. It is in the application of science, in technology, that there can be and has been a quest for power, for wealth, and for control and domination over nature and other persons.

From scientific studies can come the knowledge of our world that allows all of us to deepen our understanding of our origins and our place in nature. We still need to decide what we wish to be, to become, and more fully to develop morals and ethics. Science cannot dictate that, and cannot find answers to such issues. In contrast, the humanities can contribute directly to such issues. Science cannot tell us what should be, but it may help us to get to where we want to be.

Will it be a world where the "haves" continue to get more and more while the "have nots" continue to possess less? Will it be a world that is overpopulated with humans struggling to survive, or one where our numbers are sustainable and in harmony with the rest of nature?

Will it be a world where genetic engineering fixes genetic errors in humans and improves agriculture? Will it also be one where we attempt to engineer even "better" humans through genetic modifications? Dare we do such tinkering? If we don't are we just leaving things to chance? If we do tinker, who is to define what a "better" human is? Perhaps we will return to such questions as we discuss Huxley's *Brave New World*. This is a novel that points to a dark side of technology.

There are many areas where science and humanities need to communicate, and this clearly is the case with moral and ethical issues. We need to get beyond the battles between the two cultures that we have today, and move on to grapple with long-term issues concerning the future of humans and the future of the world.

Bowen

What a different approach David has taken on scientific ethics from his previous prognostications on scientific revolutions! On the one hand, he foresees a world in which all human thought is potentially understood in physical, scientific terms and therefore open to manipulation, and on the other, he calls on the rest of us to resolve the results of such speculation, to see that only beneficence ensues from the power produced by that knowledge. Who is going to make those decisions except governments and corporations, the first, only fleetingly interested in anything but becoming more powerful, and the second seeing profit as the all-justifying incentive?

The humanists, long regarded as speculative, nonproductive naysayers, are increasingly hard pressed to champion causes successfully that oppose the ill use of knowledge/power in today's political climate. I agree that science has gotten its own ethical house in pretty good order. But that it does so, as David pointed out, is in the spirit of enlightened self-interest, in that science has devised an internally consistent, self-governing system of checks and balances on the acquisition of knowledge. What can't be proven is outside the pale. But that includes speculation—creating plausible scenarios of the future before the facts are known. Adopting a "just-the-facts-ma'am" stance hardly equips scientists for the omniscience requisite to their increasing body of knowledge. But scientists can't avoid the responsibility that their knowledge brings them: they can't wholly divorce speculation from reality without overlooking the potential consequences of what they do.

I don't want to rehash what was already said about the relationship of knowledge and power. It is enough to say that in certain rare circumstances literature can work as a corrective, a warning regarding the effects of newly acquired knowledge and technology, and perhaps do something toward increasing the quality of life by examining human motives and behavior. That is what *Brave New World* is all about. One role of literature is to bring into perspective possibilities regarding the human condition as affected by new technology. But our speculation is chiefly hindsight to be applied to the future, instead of the almost daily accomplishments of science. When scientists are called upon to make decisions, like David's example of developing the atomic bomb, they can't be excused for being naive about the bomb's potentialities. The old idea of knowledge as evil is another way of saying that any group possessing such knowledge could acquire too much power, reshaping the structure of academe and the rest of the world in its wake. How can the disenfranchised step in to help save what's left? We try, but more often than not, in vain.

As was the case in the Manhattan Project, there are always scientists with qualms about the results of the knowledge and technology they have unearthed. But the ethical and moral visionaries seem to give ground to those who place less credence in withholding potentially devastating knowledge than in developing it. The weapons people who fell all over themselves acquiring the services of Wernher von Braun and the Peenemünde crowd[1] could hardly have been so successful if our scientists refused to work with them. The results were mixed: "One [short] step for mankind" onto the moon, and an increasingly dangerous arsenal of ICBMs for potential sale to the general terrorist public by impoverished Russians.

13

Brave New World

Bowen

Brave New World is either a utopian or dystopian novel, depending on your point of view. When I first read it as a libidinous sophomore in college in the conservative 1950s, I thought it might very well be utopian, but now in disgruntled old age, it seems more dystopian. The classification itself refers primarily to a novel usually set in the future, with circumstances and ideas current at the time of its writing pushed to their logical conclusions at some time in the future—in the case of *Brave New World*, six hundred years into the future. I thought in the fifties that Huxley had a pretty good solution to the more pressing problems I shared with my fraternity colleagues, those that dealt with the repression of sexual appetite. Today's college scene apparently enjoys a measure of utopian sexual liberation without having to sacrifice either emotion or intellect. Lest we forget, the book has been in print for nearly seventy years. What made it interesting to youthful readers in the post–World War II days is less important now than the implications of social and economic engineering and the moral and intellectual issues involved in its controlled society.

The difference between Huxley's book and its companion volume, George Orwell's *1984*, is that on the surface the sacrifice of personal liberty by the governed seems easier to live with in *Brave New World* because of the voluntary cooperation of the people affected. They can hardly react otherwise, since they have all been conditioned by genetic, chemical, and psychological mind altering, as well as being provided with palliatives like universal indiscriminate sex and soma to suppress anxieties. As Huxley tells us, the ideal ("really efficient") totalitarian state managers "control a population of slaves who do not have to be coerced, because they love their servitude" (xv).

Huxley takes for granted that his readers will be offended by the loss of individual freedom of dissent—of being dissatisfied with the status-quo—but today I am not as certain that this freedom to dissent is as important an issue for many people as it was, for instance, during the sixties. However, Huxley does not make unblemished heroes of the most discontented of the *Brave New World* citizenry, John "Savage," Bernard Marx (what's in a name?), and even Helmholtz Watson, the charming jingle-writing propagandist. Unstinting acceptance of any person, creed, or predisposition is not part of Huxley's satiric, rationalist perspective. Related to Thomas Huxley, the colleague of Charles Darwin, on one side and to Matthew Arnold, the great Victorian humanist, on the other, he spares neither science nor humanistic principles the brunt of his satire. It is important to see that John, the "savage" who is "rescued" from a New Mexico Indian reservation, is every bit as misguided in his complete reliance on Shakespeare and American Indian religion to describe human destiny as were the ruling predecessors of the Controller, Mustapha Mond, to rely on genetics and conditioning as a solution to the world's problems. In the broadest interpretation, literature and the humanities (as represented by Shakespeare) are as insane as the biologists' and social scientists' totalitarian version of a brave new world.

If John relies on Shakespeare for his framework of perception, so does Huxley himself. The book's title comes from *The Tempest*, in which Miranda, daughter of Prospero (the deposed Duke of Milan turned wizard)—after having been marooned on an island with her father and two congenitally opposite servants, Ariel and Caliban—finally sees the first European stranger (Ferdinand) in her experience with outsiders. Totally innocent of the corruption of the outside world, she confesses her attraction to Ferdinand and this unknown world of sophisticated, beautiful people in a speech that concludes, " O brave new world that hath such creatures in't" (v.i. 183–84). It is a speech of innocence, and in Shakespeare's time, as well as in the present of the novel, is grievously naive regarding the corruption and sinister motivations of the outside world that put her and her father on the island in the first place.

Building on a historical account of a shipwreck that deposited the future governor of Jamestown on the island of Bermuda, Shakespeare designed a tale of initiation for Miranda and an allegory of colonialist usurpation in Prospero's enslavement of Caliban, after disposing of Caliban's mother, Sycorax. Like John in *Brave New World*, Caliban learns from a woman, Miranda, how to read, but, unlike Caliban, John never masters cynical sophistication regarding morality, cunning, and rebellion. Like Bernard Marx, Caliban is not the most handsome of men, and his growing attachment and physical need for Miranda works toward his undoing with Prospero, ruled by old world standards of sexual morality. Linda, the mother of John "Savage," assumes the role of the wronged and pugnacious Sycorax, who is governed by a cultural conditioning far removed from Prospero's regime. In the novel Linda is more sinned against than sinning, more victimized than even her son by both the Pueblo Indian and the brave new world civilizations.

In *The Tempest*, Ferdinand, ritually reborn in the sea like the rest of the passengers and crew wrecked by the tempest Prospero has brought upon their ship, emerges from baptismal waters of forgetfulness, and ultimately their collective salvation is restored through their new knowledge of life and morality under Prospero's magic tutelage. Huxley's novel creates a diametrically opposed message, as he applies the story to the eighteenth- and nineteenth-century concept of the "noble savage," the unspoiled innocent raised in a primitive environment free of the taints of life in "civilized society." However, in Huxley's novel John is contaminated by the beauty of Shakespeare's seductive, romantic, pain-insistent poetry and driven to the point of a suicidal response to the irreconcilability of the two cultures.

For John it is not merely a choice of clichés: "No pain, no gain" (for him, Christian redemption through suffering) or "Don't fight City Hall" (the rulers of the brave new world), but a more subtle one between the border-line masochism of spiritual gratification afforded by sexual deprivation on one hand, and hedonistic pleasure at the cost of mind-numbing surrender of both personal responsibility and the freedom for individual thought and actions on the other. Huxley concedes in his introduction to more recent editions of his book that such a contrived and limited choice was neither realistic nor aesthetically honest. Joining Helmholtz and Bernard on distant islands where dissidents were sent in order to pursue their own individual inclinations seemed a cold, but comfortable enough, compromise for John to make. But such a choice, as Huxley tells us in his introduction, would have implied a rational sanity that John's devotion to

Shakespeare would not have allowed. He cherishes suffering, being educated to a Renaissance martyrdom that took all the worst of a Christian conscience, tempered it with what Huxley regarded as Renaissance irrationality regarding the perfectibility of mankind, and made it impossible for John's hybrid personality to exist free of the encroaching science-dominated cultural atmosphere he could not avoid.

The science that drives the brave new world is of a particular kind, as Huxley tells us in the foreword:

> The theme of *Brave New World* is not the advancement of science as such; it is the advancement of science as it affects human individuals. . . . The only scientific advances to be specifically described are those involving the application to human beings of the results of future research in biology, physiology and psychology. (xi)

The devices of the genetic and psychological sciences that were applied in the novel to the development of normal human behavior were certainly within the scope of possibility when Huxley wrote the book. He exploited existing practices and applications pushed to foreseeable ends by the rulers, while the philosophy of creating workers ideally suited to their tasks goes back at least to Plato's *Republic*. In the nineteenth century, ideas like genetic classes of superior beings became ever more accepted as a part of the propaganda of imperialist rationalization for overcoming "inferior," "native" beings and races all over the developing world, from Nietzsche's idea of the Übermensch to British anthropological studies and skull measurements, mentioned earlier. Malthus and others introduced grand schemes of social engineering, but the problem was, especially after the French and Russian revolutions, to keep the lower classes happy and productive. That problem, as Huxley tells us, remains to the present day. "It is the task assigned, in present-day totalitarian states, to ministries of propaganda, newspaper editors and schoolteachers" (xv). But as we have seen in our Foucault discussion, a lot of people remain uncooperative. It is only a short stretch of the imagination to see how the biological, social, and psychological sciences could be pressed into the service of economic stability. We were already well on the way to differentiating "normal" from "abnormal" behavior, and utilizing the scientific resources of a drug-sanctioned psychotherapeutic culture to calm unruly people. While genetic engineering is only now coming into full bloom, the very arguments Huxley predicted for six hundred years hence are strangely prophetic of current debate, as Skinner's behavior models still obtain in many American universities today.

Huxley's structure reiterates his themes. We are immediately introduced to the novel by visiting the "CENTRAL LONDON HATCHERY AND CONDITIONING CENTRE" (1). The title of the human factory hangs under the motto of the World State, "COMMUNITY, IDENTITY, STABILITY" (1). These signs over the door encapsulate the means and ends of the brave new world: the interweaving of genetically engineered conditioned responses in humans to erase individual identity and produce economic (and therefore social) stability. Reminiscent of David's Fourth Revolution of Science (I can't resist stretching a point of comparison here), the reproductive process is mechanized, beginning with the separation from the human body of sperm, ovum, embryo, and fetus, and continuing with their development along one continuous assembly line, where chemical and psychological checks are maintained. The assembly line process, first used on a grand scale by Henry Ford, is integral to the transference of Christian ritual, symbology, and iconography to the deification of Ford.

This mechanistic brave new world is described in direct contrast to the "other," "primitive" world in which John "Savage" has been raised, which doesn't seem much better. Huxley's description of the "savage" hybrid Christian/Indian ritual (snakes, painted eagles, crucified men, and torture)(114–16) sounds as if it too had been subsumed into the orgiastic brave-new-world solidarity rites, except that the original scapegoat/sacrifice of the primitives has been given up in the brave-new-world version, and the sexual impulse underlying religious frenzy expanded to its normal copulative conclusion.

Huxley apparently didn't put much stock in religious practices. Sayings like "Ford's in his Flivver" and "Thank Ford" act as comic puns while they work to demonstrate a world run not by introspection as much as slogans, which, we are constantly reminded, are the products of early multiple reiterations during each youth's sleep conditioning. As Bernard Marx tells us, "Sixty-two thousand four hundred repetitions make one truth. Idiots!" (47).

Huxley adapts the reiteration process to the structure of chapter 3 of his novel as he juxtaposes the instructional lessons of the Director and then of Mustapha Mond on the conditioning procedure with its reassured results on the victimized populace. Beginning with longer remarks and descriptions to set the scene, slogans and rationalizations are repeated again and again against a background of the ongoing thoughts and conversations of various characters. The voices of conditioned consumer/ citizens like Fanny and Lenina, alternating with that of dissenter Bernard Marx, are juxtaposed against the official party line described by the rulers in charge in a kind of repetitive fugal composition on the conditioning process itself (28–56). The chapter ends in the quiet beatific resolution of Christ via the Controller's snippet of Christian cliché, "Suffer little children." The fragment opens the possibility of opposite interpretations of *"suffer,"* and forms a coda of metaphoric immutability concluding the chapter: "Slowly, majestically, with a faint humming of machinery, the Conveyors moved forward, thirty-three centimetres an hour. In the red darkness glinted innumerable rubies" (56).

A corollary to the God/Ford identification is Huxley's further conflation of Henry Ford with Sigmund Freud:

> Our Ford—or Our Freud, as, for some inscrutable reason, he chose to call himself whenever he spoke of psychological matters—Our Freud had been the first to reveal the appalling dangers of family life. The world was full of fathers—was therefore full of misery; full of mothers—therefore of every kind of perversion from sadism to chastity; full of brothers, sisters, uncles, aunts—full of madness and suicide. (38)

There are grains of satiric truth to all this. Freud asserted that in some cases traditional family relationships were challenged by an underlying subconscious defying taboos about incestuous relationships. These sexual taboos were nullified by the state's elimination of families and instead by enhancing the sexual impulse behind uninhibited religious worship when Sunday services were replaced by "Solidarity Days" ending in orgiastic copulation.

Huxley made other ingenious conversions from Christian to Fordian symbolism, such as raising the bar on the Christian cross to produce the T symbol for Ford's model T. Such verbal and iconographic shenanigans indicate Huxley's vigorous cynicism regarding the original iconography and its meaning, subject to easy assimilation into new trademarks, slogans, and associations. Huxley plays on Ford's near deification in the early twentieth century, first by capitalists for his new economics of mass production, which made Ford's flivvers, Model A's, and Model T's available to most working people, and at the same time dramatically increased profits for Ford; and second by workers for his $5.00 a day minimum wage, allowing them to purchase the Model T fruit of their labors. Ford's assembly line was regarded as a panacea for human consumption as well as the means of economic salvation. Thus, the human biological assembly line is the epitome of the brave-new-world economy. The economies of mass production are why the Director of Hatcheries and Conditioning is so excited in his explanation of how many hatchlings they can get out of the same ovum, when uniformity and conformity are especially desirable among the lower working classes, whose jobs require little individual decision making.

Thus, the closer people get to being entirely controllable machinery the more efficient the economy of the world becomes. One of the recurring themes that has surfaced during our discussion is how the "hard" sciences stick to chemical and physical situations they can observe and predict through experimentation, while the humanities seem apparently stymied in their lack of progress toward "laws" and "truths" that would promise the same certitude. So Western society has adopted quasi-scientific speculation as a means of "normalizing" the human situation. Instead of celebrating individuality, we tend to demonize the "other" in terms that denigrate any difference from ourselves or the group with whom we have been conditioned to identify—politics, nationalism, race, skin pigment, sexual preference, religion or lack of it, and so on. Having been conditioned to an acceptable definition of "normalcy" we either take up arms against the "abnormal," or—more humanely perhaps—try to convert the "abnormal" to "normal" or redefine our understanding of "normal" to comprehend the anomalies. For example, the declaration that homosexuality is no longer to be considered a medical disorder to be treated has become more widely accepted in recent years.

In our own world as well as Huxley's brave new one, conditioning is everything. That is why the brave new world process continues beyond the Hatchery and into childhood. The abolition from polite conversation of the terms and activities of motherhood, and their relegation to the language of profanity are all part of the conditioning process. Conversely, our current insistence on "family values" is to award ultimate credence to the prejudices, ignorance, or stupidity of parents and other family members over other social conditioning factors. While we may be justified in supposing any random set of parents to be no more intellectually or morally capable of conditioning their offspring than anyone else, there are also favorable odds of a genetic as well as a conditioned response in a mother's nurturing of her child. Again the evil in *Brave New World* lies in the assumption of the rulers that the conditioning should first be of benefit to the state rather than the conditioned individual, when we suspect, especially after reading Foucault, that the total identification of the good of the state with that of the individual can often prove false. Choices made by individuals in *Brave New World*, except those made by the handful of dissidents, are not informed by understanding, but by conditioning. In the novel even alpha class people are only given the information necessary to their preordained tasks in life. The lower the class, the simpler the task, and the less information provided about the whys and wherefores of how anything works. Everything, as the latest CIA movies inform us, is on a "need-to-know basis." Knowledge breeds informed reasoning and dissent.

Because Huxley has to provide his readers with a greater understanding of the rationale for the whole world program, he puts that explanation into the final conversation between the book's dissidents and the Controller, Mustapha Mond, originally a physicist, who is conversant with forbidden texts and a would-be dissident himself. Mond has made the choice to join the system/plan rather than attempt to beat it, and comes as close to being persuasive regarding its rationale as anyone in power could be. In such a world, exile, along with other presumably intellectual dissenters, seems a less odious choice than reform or residing in a world of slogans and conditioned response.

Isolation represents a traditional recourse for an intellectual living in any world he can't respect or influence, from the days of Voltaire's Candide, alone and hoeing his own garden, to the present. But it is not for everybody, as the Controller explains:

Mother, monogamy, romance. . . . The urge has but a single outlet. My love, my baby. No wonder these poor pre-moderns were mad and wicked and miserable. Their world didn't allow them to take things easily, didn't allow them to be sane, virtuous, happy. What with mothers and lovers, what with the prohibitions they were not conditioned to obey, what with the temptations and the lonely remorses, what with all the diseases and the endlessly isolating pain, what with the uncertainties and the poverty—they were forced to feel strongly. And feeling strongly (and strongly, what is more, in solitude, in hopelessly individual isolation), how could they be stable? (41)

The disaffected of the book, individualists for various reasons, Bernard, Helmholtz, John, and to an extent, even Mustapha Mond, find companionship in their questioning an order that eschews "otherness." While Bernard's disaffection stems from his feeling of being rejected, the handsome Helmholtz writes a poem (comic for some of us) about the sanctuary solitude offers from continual meaningless sexual activity with a host of adoring women, while John seeks solitude from his own conditioned notions of love's sanctity disintegrating with the fire of his own lust and that of the ever-willing object of his worship.

The sinister aspects of total gratification are apparent in the death of John's mother, Linda, a victim of sensual satiation from the drug of brave-new-world choice, soma. She dies in blissful oblivion in a hospital that is also a center where children are conditioned to ignore any sentiment connected with death, normally the greatest deprivation of all. Her situation is the trigger for the open rebellion of her son against the whole situation. As John leaves the hospital full of the agony of death, he encounters the daily soma distribution to the Delta workers, and the full irony of Miranda's declaration in *The Tempest* dawns on him: "'How many goodly creatures are there here!'" The singing words mocked him derisively. "'How beauteous mankind is! O brave new world . . .'" (215). As the 162 twinned Deltas line up like zombies for their soma, Miranda's words are repeated, but

Now, suddenly, they trumpeted a call to arms. "O brave new world!" Miranda was proclaiming the possibility of loveliness, the possibility of transforming even the nightmare into something fine and noble. "O brave new world!" It was a challenge, a command. (216)

Opposite to the recourse of isolating oneself from the encroachments of the new society is (echoing Hamlet) to take up arms against the sea of troubles and, by opposing, end them. John's quixotic attempt to lead the Deltas to full-scale emancipation from the system, fruitless as its seems, does cause an unwilling Bernard and a more willing Helmholtz to seek some solution to their problems with the system, and ultimately to obtain the conclusive informative interview with the Controller on the realistic—rather than propagandistic—basis of the brave-new-world society, and finally get a genuine philosophical choice as to their future course of action.

Chapters 16 and 17 comprise the final obligatory interview with the Controller. Now acting the Prospero role as all-seeing magician responsible for the brave new world, Mond lays out its entire societal plan and justifies it in both practical and philosophical terms for both the dissidents and the reader. Resembling a similar scene in Orwell's *1984*, it represents the novel's present dystopia as a consequence of humanity's own folly. The world has abandoned the artifacts of high art, literary and otherwise, for the latest products of the commodity culture, which depends upon a happiness born of the satiation of immediate wants rather than any appreciation of things that are unsettling, tragic, or even thought-provoking—all of which are socially destabilizing:

> We don't want to change. Every change is a menace to stability. That's another reason why we're so chary of applying new inventions. Every discovery in pure science is potentially subversive; even science must sometimes be treated as a possible enemy. Yes, even science. . . . It isn't only art that's incompatible with happiness; it's also science. Science is dangerous; we have to keep it most carefully chained and muzzled. . . . science is a public danger. As dangerous as it's been beneficent. It has given us the stablest equilibrium in history. . . . Our Ford himself did a great deal to shift the emphasis from truth and beauty to comfort and happiness. Mass production demanded the shift. Universal happiness keeps the wheels steadily turning; truth and beauty can't. (231–34)

Chapter 17 completes the rationalization of the new order, dealing primarily with the Controller's assessment of God's place in the new world, and ultimately coming to the conclusion that "One believes in things because one has been conditioned to believe them. . . . People believe in God because they have been conditioned to believe in God" (241).

In a sense the indignant young Huxley wrote *Brave New World* in response to what he saw to be the dangers of a totalitarian if benign society, scientifically limiting the independence of the populace to question and possibly disrupt the smooth operation of the state. But nowadays even "the state" is increasingly giving way to "the global economy," linked by Internet ties, which lumps all together in one enormous consumer culture. Not even Huxley's prophetic book has slowed the conditioning of mind-numbing propaganda and stupidity bombarding the populace every day. As I have mentioned in earlier chapters, the corporate conditioning by advertisers spouting slogans in the media increasingly assails our senses every day. Our political and social institutions are corrupted by twisted TV slogan-"truths," repeated so often they assume the guise of reality, so that no one can hope to be elected to office without the bribe money needed to purchase it. In the late 1950s Vance Packard's book on subliminal conditioning, *The Hidden Persuaders,* came as a revelation, but now we are so conditioned that the practice barely warrants notice. In this respect we are not so far removed from Huxley's projection.

The major harbinger of *Brave New World*'s commodity culture that has come to pass in the present is its repeated insistence on promoting consumption :

Imagine the folly of allowing people to play elaborate games which do nothing whatever to increase consumption. It's madness. Nowadays the Controllers won't approve of any new game unless it can be shown that it requires at least as much apparatus as the most complicated of existing games. (30)

The living rooms of my grandchildren's homes are a cluttered testimonial to this philosophy. The bathrooms of our house contain disposable razors and tissue paper nose wipers; our kitchen shelves contain throwaway cameras and paper plates; our libraries are replete with complicated CD-ROM games; and our grandchildren own a dozen video cassettes starring a mindlessly happy dinosaur—the list goes on and on.

One frightening aspect of the novel is the brevity of Huxley's historical chronology of how the brave new world got that way:

Then Nine Years' War began in A.F. 141. . . . The noise of fourteen thousand aeroplanes advancing in open order. . . . the explosion of the anthrax bombs is hardly louder than the popping of a paper bag. . . . The Russian technique for infecting water supplies was particu-

larly ingenious. . . . The Nine Years' War, the great Economic Collapse. There was a choice between World Control and destruction.
Between stability and. . . . (47–48)

Huxley lets us fill in the details. Since World War II we have thought of
Armageddon so often he doesn't need to do more than hint at the scenario. While he didn't recognize the potential of nuclear weapons in the
1930s, we did manufacture anthrax bombs and still have stockpiles in
storage. But even Huxley could not have predicted all the ways that scientific advancement could threaten our very existence. In the pre-DNA
days he could only suspect how close we would come in the twentieth
century to cloning of humans, genetic tinkering, and the host of potential
threats to individualism that current biological sciences have made a distinct possibility. But Huxley does offer some rather cold comfort. As an
alternative to our present propensity for holocaust by weapons of mass
destruction, the brave new world doesn't sound so bad.

John's ultimate retreat to a remote lighthouse to indulge himself in
masochistic punishment fantasies is interrupted only momentarily by
the pleasure he takes in making a bow and arrows, before his singing
reminds him of his sinful enjoyment. In the concluding metaphor of the
book, his self-flagellation is caught on film by an enterprising photogra-
pher and his seemingly insane difference from "normal" civilians is exposed again. As a consequence he resumes the role of carnival freak for
the unremitting curiosity of the populace. The brave new world, unsuited
to his propensity to suffer, becomes impossible for him to inhabit, so,
choosing the ultimate isolation, he hangs himself in his lighthouse.

While science may or may not warn us of the dangers inherent in our
march to conquer the secrets of the universe, it is up to the artists and
fiction writers to narrativize the possibilities of such power in such a way
that we can come to a human rather than a mechanical formulation of
what could be a clear and present danger—to realize its potential in metaphoric terms immediate to our experience. The creative imagination, for
all its lack of verifying checks and balances, sometimes acts as both a
warning of the adverse possibilities of taking certain paths and a deter-
rent to the scientific hubris that could put us on those paths.

Response from Wilson

Aldous Huxley wrote *Brave New World* between World War I and World War II. He wrote a new foreword for the book just after the latter war. This was right after the atomic bombs had been dropped on Hiroshima and Nagasaki, and, in his foreword, Huxley predicted dramatic changes in society as a result of the future development of nuclear energy:

> Nuclear energy will be harnessed to industrial uses. The result, pretty obviously, will be a series of economic and social changes unprecedented in rapidity and completeness. All the existing patterns of human life will be disrupted. . . . The nuclear scientist will prepare the bed on which mankind must lie. . . . These far from painless operations will be directed by highly centralized totalitarian governments. Inevitably so. (xii–xiv)

These predictions of Huxley proved wrong, which may say something about his, or anyone's, ability to predict the future. Determining the effect that a new technology will have on the world is a most difficult thing to do. We all can be thankful that the dire predictions of many concerning nuclear weapons have not materialized, at least so far.

Nevertheless, I strongly agree with Zack that novels, and today's films, can contribute to our attempts to foresee, and hopefully avoid, some of the less attractive aspects of technological advances. From books such as *Brave New World* and *1984* to the more recent films, *Road Warrior* and *GATTACA*, the arts have ways of influencing ideas and our futures. But, as with science, such influences can be positive or negative, and outcomes are not always under the control of the inventor.

14

Summary

Wilson

Scientific methodology has given us a reasonably accurate view of the world. It is never a final view, and we can never know for sure that it is a true view, but we know certain things about the world beyond reasonable doubt.

The indirectness of our sensory input does not seem to be as limiting as it might be: we can discover new phenomena that we cannot sense, such as x-rays and radioactive decay, as instruments extend our senses. We can correct errors, such as those of optical illusions. Our imaginations, coupled with science, enable us to build models of reality.

Such models are not perfect, but seem to be getting better. There is progress in science. Even with the possibility of short-term detours and dead ends, our theories and models encompass and explain more and more data about the world. This continues to be true even as revolutions change some of the ways we do science.

Though the meaning of words is not precise, nor even exactly the same for different individuals, there is enough of a common understanding to allow science to proceed. Furthermore, the ability of words to change meaning with time is an advantage for science, since the meanings can develop with our understanding of reality.

The strength of science can be seen especially clearly when different sciences, working independently, come to similar conclusions about shared issues. The fact that astronomy, geological science, and biological sciences all have arrived at consistent dates for the age of the earth, and the time of origin of life on earth, strengthen those estimates. Molecular biology, with its recently developed ability to trace mutational differences in the DNA of different organisms, generally confirms the tax-

onomy of organisms worked out on the basis of their physical characteristics, and both confirm the sequence of appearance of life forms as they
are found in the fossil record, greatly strengthening the theory of evolution.

Although the success of science is reflected in the development of new
technologies, the limits to our scientific knowledge are also reflected in
the unforeseen consequences of some of these technologies. We all share
responsibility for these technologies. We all are involved, especially in a
democracy, for the decisions that are made and the products that are
bought. We also elect the presidents, senators, and others, who decide to
develop the weapons and when to use them.

While the practice of science is not perfect, and scientists are not perfect, the community of scientists working together has allowed for
progress. Individual scientists carry biases and a small minority are not
truthful about their work, but these flaws do not seem to do irreparable
harm to the enterprise as a whole. There is always a risk of distortions
emanating from self- interest, as well as a risk of one-sided development
of knowledge due to the special interests of those who supply the funding for so much of today's scientific research. However, there are self-
correcting mechanisms in science, such as the repetition of experiments
by others and the drive to develop theories that best fit the data, even if
revolutionary change is required.

Thus, risk comes not so much from science that is wrong, but from the
lack of knowledge we suffer from in under-supported areas of science.
The power to control what studies are funded is, in the end, the power to
control what kinds of things are known. Social and governmental decisions will then be based on partial knowledge, which could result in actions whose consequences will not be fully understood because of what
we don't know. Will we know enough about ecology to realize the risks
of the introduction of a particular new chemical? Will we know enough
about how to make cheap solar energy before running out of coal and oil
if we mostly invest research dollars in fossil fuel exploitation?

Scientists and technologists have special moral obligations as they decide what to work on, confront issues related to the design of experiments, and attempt to foresee the consequences of their efforts. But it is
not always possible to see all of the consequences. When chlorofluro-
carbons were first used in spray cans, no one should have been expected

to predict that they would turn out to be destructive to the ozone layer of our atmosphere.

All the more reason then for us to act rapidly and responsibly when the adverse effects of technology are recognized. Air pollution, water pollution, global warming, and ozone layer depletion are but a few of our human-caused problems. We are drastically reducing biodiversity today, as species go extinct, just because of human numbers and actions. We need to get our population growth under control if we are to avoid much pain and suffering in the world in the future.

Focusing on the subject of our book, one of our main goals has been to look at issues bordering on science and the humanities. In 1959, C. P. Snow wrote a little book called *The Two Cultures*. In it he lamented the growing divide between science and humanities.

> Between the two a gulf of mutual incomprehension—sometimes . . . hostility and dislike, but most of all lack of understanding. . . . Scientists . . . shallowly optimistic, unaware of man's condition. . . . Literary . . . intellectuals . . .lacking in foresight, . . . in a deep sense anti-intellectual. (4)

The passage of time has only confirmed Snow's lament. The gap between science and humanities remains as great or may be greater today. It is essential that this gap be bridged, and we hope that our book contributes to this in a small way. We must educate the next generation to be knowledgeable about both conglomerations of disciplines and how they interrelate.

Science is rapidly advancing our knowledge of the universe, and building from science, technology is enhancing our power to alter the world and civilization, for good or for bad. At the same time, science cannot serve as a source of human goals, nor can it be used to distinguish right from wrong. There are insights into human nature that come from the arts and humanities. As many of us as possible need to know enough about science and enough about humanities to make intelligent decisions concerning where we are headed and what we wish to be, as individuals and as a civilization.

I have suggested that a revolution is under way that could greatly influence the way we view science, humanities, and ourselves. That revolution has to do with the relationship between mind and brain. To the extent that mind, consciousness, and our experiences arise solely from the activities of the brain, all of our disciplines are tied together—from science through the humanities, as E. O. Wilson has pointed out in his book, *Consilience*. The linkages between humanities and science are as real as the synaptic connections between brain neurons. There may be no insurmountable barrier between the social and natural sciences, and ultimately there may be no such barrier between science, on the one hand, and arts and humanities on the other.

During our explorations in this book, I hope that you have gained a little insight into these issues and perhaps into your own nature and purpose. Zack and I have left you with challenges: the nature of free will; the possibility of success in making the social sciences truly scientific; the problem of unforeseen moral consequences of the technological applications of scientific discoveries, and others.

In *The French Lieutenant's Woman* we looked at issues like power and politics, and the application of scientific laws to humans; we explored the role of chance and uncertainty in human affairs in *The Crying of Lot 49*; we considered ethics, morality, and the potential for misuse of technology in *Brave New World;* and we cautioned against the possibility of science becoming a religion of its own.

Finally, I hope that Zack and I have been enough of a Laurel and Hardy team to have kept some humor in as we discussed these serious matters.

Bowen

As I see it, our book has aimed both to delineate the ways in which literary and scientific people regard the world and to show some overlap among the questions they consider and the ways they express those considerations. Often we have broadened our frame of reference beyond our immediate areas of study and research to comprehend three vast areas of human knowledge: science; arts and humanities; and the hybrid disciplines that incorporate and cross over the boundaries that classically define the other two: namely, economics, the social and political, psychological, cultural and anthropological sciences, and history. These hybrids, which often gain status from their affiliation with "hard science"— chemistry, physics, biology, and geology—have recently added geography, as that hybridized form has turned its attention to such scientific matters as oceanography, and cultural/social issues such as urban geography and economics.

Offshoots of all disciplines are in a state of constant evolution, as boundaries and perspectives of various subgroups draw on the methodologies and ideas of their predecessors. Even the hard sciences regularly spawn new disciplinary offspring, realigning their focuses, chasing the research money in different ways. A quick telephone-directory count of departments in the University of Miami Medical School reveals approximately 240 different departments. Even discounting a third of the listings as paper-shuffling and billing mechanisms would leave nearly 200 subdisciplines represented in that institution alone. The number of departments with the prefix *bio-* is particularly large.

In many universities with technological reputations, the College of Engineering includes a lot of departments housed elsewhere at our school. The two schools nearest in size to arts and sciences here are Business and Communications, both once departments and now mini-empires on their own. You would be hard pressed to find many universities or colleges in the country that align the subject matter of what is taught in similar departmental patterns, but the national tendency seems to be an explosion of areas of expertise. For example, the Department of Philosophy offers various courses in the philosophy of science, while statistical analysis is up for grabs among rival competitors from Mathematics, Computer Science, Business, Statistics, Physics, and possibly others.

The result is that human knowledge as we think of it is increasingly fragmented, even as it crosses over from discipline to discipline in pursuit of tuition and research dollars, and in response to new ways of looking at things. So if there is no agreement on the proper division of subject matter in the various departments and disciplines in universities as a whole, who among us is to make any such arbitrary classification? Disciplinary information has reached entropic proportion. Yet scientists do talk to each other and they are heard and read by pseudoscientists and humanists alike, all of whom give their idiosyncratic spins to new discoveries, innovations, trends, and revelations.

Part of why I am here is to provide some kind of perspective on what those scientific ideas might mean to intelligent people at the other end of the disciplinary spectrum. As literature professors, I and my breed try to deal with the subject matter of all things that interest us for political, economic, social, or aesthetic reasons. While we were originally aestheticians in our approach, we also have been expanding into subdisciplines that undertake to comprehend not only how the words are put together, but what they mean, or more recently, whether, as abstract paradigms or artificial verbal imitations, they have any meaning at all, even as we use their own expressions to express our disbelief.

Literature professors are also psychoanalysts as they try to understand the meaning of words in terms of the writer's hidden motives, subliminal agendas, and politics, and we are often moralists as we see messages, hidden meaning, conspiracies, and social agendas in the works we read and interpret. Thus, we are as idiosyncratic a group as any in other disciplines. But we are painfully conscious of our differences. We increasingly regard literature in terms of the reader's response to it, the anxieties of the writer writing it, and the publisher's motives for publishing it. The one thing that most of us distrust is the certitude with which those of various credos espouse their own version of the truth. Certitude itself has become a cause for doubt, even as we offer a certitude about certitude itself. As Yeats says, "The best lack all conviction/While the worst are full of passionate intensity." From this distrust of absolute truth springs a disrespect for any proclaimed truth we assume to be born of an agenda tainted, we suspect, with the politics of cultural ideology.

Assuming science to be the dominant ideology of the twentieth century and certainly destined to be dominant in the twenty-first, I have tried to indicate how science is incorporated into a larger scheme of ideologies, political, religious, moral, and social, and to view it as simply another construct among the multitudes concocted by man since the beginnings of recorded history. I have attempted to demonstrate that the hierarchies and paradigms of science resemble those of religions, mythologies, and creative ideas far from the particular discipline that science has imposed on itself: empirical data arrived at through controlled experiment, and replicable in other labs and places producing similar conclusions. If the method of discovering such "truths" is unique to science, what the apparatus has become in social, political, and economic terms is certainly not. People have simply put science at the top of an ever-evolving hierarchical order. Two things, as I see them, have resulted from this hegemonic ascendancy. Once the search for pure truth seems to pay off in terms of power, pure scientists either abandon the consequences to technologists or politicians, or they assume the prerogatives of philosopher-kings in touch with the forces and realities of the universe, like Professor Rothman, angry with others who, lacking their scientific rigor, or methodological approach, question or propose alternatives to ultimate scientific truths.

One science, physics, is closest to the realm of universal absolutes outside the laboratory. The theologians of the scientific community, they are after the eternal hypotheses regarding matter and energy that affect their more immediate lab-bound buddies. We have concentrated on those big physical truths—the tables of the law—to put it in biblical terms, and ultimately to apply them to all life and inorganic matter. With the imminent conjunction of the biosciences and physical laws, we have brought the subject directly into the domain of the humanities: their interaction with people, the intrusions on human behavior, the potential control of freedoms, trials, tribulations, ecstasies of the human mind, and the construct we call the human spirit.

Complementing the hope of understanding and eventually gaining such control of human destiny as scientific knowledge will allow is science's professed ability to deal with both human and inorganic history. The application of scientific methodology to such questions as the origin and history of the cosmos, treating these as amoral, nontheological, physical questions, equates with an infant scientific belief system disregarding as worthless the history of how such things have been dealt with through thousands of years of recorded history.

While I have as great a belief in the myth/belief system that is scientific methodology as anyone, I have tried to temper its arrogance as I would the belief in any system claiming to have, or be in the process of acquiring, the absolute truth about anything. But I never doubt the ingeniousness of the process, perhaps as great or greater than any that drove the elaborate civilizations of the past. Scientific progress through its offshoot, technology, has brought about things undreamed of, and no doubt will continue to do so. Arguably, technological progress has made life, considered by Hobbes "mean, brutish, and short," a little longer, but I worry about the "mean and brutish" part. Thanks to science, life could indeed be made more comfortable for everybody, although to date it apparently hasn't done so for a lot of people; and less bestial for everybody, although it doesn't seem to have done that either. Is it fair to ask one belief system to answer all human and environmental problems? Of course it is if the system promises ultimate knowledge of the universe, past, present and future, and solely through its agency.

But even while questioning the absolutist role of science, I must agree that it is nevertheless one of the most creative systems to come along in the past two hundred years. It has caused major reorganization of other systems—from law and medicine to arts and literature. New art forms are the offspring of science, from new concepts in architecture to new substances for the plastic arts, to motion pictures and even television. Literature draws its subject matter from rethinking the old culture tempered by a new one identified with science. It is impossible to read any post-fifteenth-century book not directly or indirectly involved with science.

In this book I have tried to introduce literature which directly addresses scientific phenomena to add an even greater creative dimension to both scientific ideas and the idea of science itself. The great thing about fiction and poetry is that they don't derive their truth from scientific verifiability, but, as Blanche Dubois says, "what ought to be true," or what might possibly be true, or what could be true if. . . .[1]

Appendix A

Science and Literature—The Course

For teachers interested in ideas about how such apparently divergent material might be approached in an upper-level undergraduate curriculum, we provide the following mechanics of the joint course "Science and Literature" which we taught during the spring semester of 1998 at the University of Miami.

The class took the course for senior credit in either Biology or Literature, depending on their need for major requirements. The result was intended to be a capstone experience for the students, who were forced to bridge the two cultures. We began each topic with background information on those scientific concepts necessary for the analysis of selected literary works. That analysis led to more compelling issues at the interface between science and the humanities. Dialogues between the two of us opened the way to discussion sessions among the students. The interchanges among instructors and students alike forced all of us to think more deeply.

The mechanics we used to merge two classes into one probably would work for many pairs of classes at most institutions. Each of our versions of the course was listed separately in the schedule with an explanation of how it was going to work. With the merging of what would otherwise have been two classes of students into one, there was no real loss of teaching load, even with two of us together in one classroom for each session. Thus any prospective bean counters would not have been able—even if they hadn't been as willing as they were to have the course take place—to argue that there was an extra cost to the university just because the arrangement was team-taught. Enrollments were good in both English and biology, and student satisfaction was amply demonstrated in their course evaluations, so at least the class apparently came away satisfied with their learning experience. When we represented the University of Miami at a conference on "learning communities," the response of people from other institutions was so positive and the requests for more detailed information so apparently genuine that we felt justified in our decision to develop and disseminate our ideas in book form. An abbreviated syllabus for the course follows.

Biology 474 Special Studies (Science and Literature); English 495 Literature and Science

(Taught together: 3 credits; Instructors: Zack Bowen and David Wilson)

Description

Views of reality and natural phenomena from science and from literature/humanities will be examined and compared. Topics will include origins of the universe and of humans, scientific methodology and revolutions, and issues related to perception, experience, language, free will, indeterminacy, cause and effect, scientific progress, and ethics.

Prerequisites

Junior standing and a major in English OR in the sciences (biology, etc.). This course counts for credit as a major's course in either English (ENG 495) or Biology (BIL 474).

Textbooks

The Science Gap, Rothman
The French Lieutenant's Woman, Fowles
Brave New World, Huxley
The Crying of Lot 49, Pynchon
Additional shorter reading selections on reserve in library.
For those with weaker science backgrounds, *The Sciences: An Integrated Approach*, Trefil & Hazen, is recommended reading.

Lecture topics and reading assignments (L = library reserve)

Date Topic

Jan 13 Introduction; Origins; Scientific world view
  Readings: Trefil and Hazen

Jan 15 Origins, cont.: Metaphors, Narrative Creations
  Readings: from Genesis and "Night-Sea Journey" (L)

Jan 20 Scientific methodology

Jan 22 Scientific laws, application of laws to human biology, history, and origin of species

Jan 27 Knowledge and Power: history and politics in science and the humanities; Social sciences: mixing experiment, statistics, and narrative
  Readings: Foucault *Knowledge/Power* (L); Tref. & Hazen

Jan 29 *The French Lieutenant's Woman*

Feb 3 *The French Lieutenant's Woman*

Feb 5 Language, uncertainty, cause and effect, and bias
  Reading: Derrida

Feb 10 Indeterminacy, uncertainty, cause and effect, chaos, determinism, free will, and bias

Feb 12 1st and 2nd Laws of Thermodynamics; Entropy, Maxwell's Demon, Information, and Computers

Feb 17 *The Crying of Lot 49:* Indeterminacy, probability factors, history, paranoia, chaos theory, and thermodynamics

Feb 19 same as above

Feb 24 First Paper Due. Review session for mid-term exam

Feb 26 Mid-Term Exam

Mar 3 Truth, Reality and Representation; Plato's cave and indirectness of sensory input

Mar 5 Revolutions in science and the nature of scientific progress

Mar 10, 12 Spring Break

Mar 17 Predictability, patterns, cycles; creativity and the notion of progress
  Reading: Yeats (L)

Mar 19 Ethics, morality; science vs. technology

Mar 24 *Brave New World*

Mar 26 *Brave New World*

Mar 31 *Science Gap*, chaps 1–4, 6

Apr 2 *Science Gap,* chaps 7,8,10,11

Apr 7 *Science Gap,* chaps 12–15

Apr 9 *Science Gap;* Open discussion with students

Apr 14 " " " "

Apr 16 " " " "

Apr 21 Second Paper Due. Open discussions continue

Apr 24 Review session for final exam

Final Exam: April 30, 2:00 P.M.

Appendix B

The Science Gap

Bowen

In our original course we concluded with Milton A. Rothman's *The Science Gap: Dispelling the Myths and Understanding the Reality of Science,* chosen for a general class discussion of his defense of science and scientific methodology. In treating Rothman's book at such length, we did not mean to elevate its relative importance, but to use it as an opportunity to recapitulate positions we have taken in previous chapters, as well as to raise other issues related to bridging the two cultures, even as it places all the issues squarely in the current popular culture "Science Wars" debate.

Rothman's book was chosen for its polemical defense of science and scientific methodology. Rothman's conception of science is an evolved but conservative derivation of ideas espoused by the Vienna Circle, a group of positivist philosophers in the 1930s and 1940s, who held that science is a strictly logical process: "Scientists propose theories on the basis of inductive logic, and confirm or refute them by experimental test of predictions derived from the theory."[1]

The Science Gap is written, according to its cover blurb, to set "the record straight" regarding the "myths" that have come to surround human understanding, particularly of science and matters to which scientific methodology applies itself. Written for "the educated lay reader," the book is characterized by a style that should be clear to its generalized audience, and in polemical contradistinction to the array of recent attacks on science by humanist theoreticians, some of whose observations we have already noted in this course. Without Rothman's mentioning any of his antagonists' names, his remarks seem to me to be a point by point refutation of Paul Feyerabend's popular anti–"Big Science" books, *Against Method* and *Science in a Free Society*.[2] In particular, Rothman's obsession with the "myths" seems directly related to Feyerabend's allegations about scientific laws, and his espousal of "intuition" and parapsychological phenomena and beliefs. Rothman's ideas provide a venue for us to recapitulate and extend some of the positions we have already discussed.

Rothman begins the book with an epigraph defining his principal term: "*Myth* n. Any fictitious story, or unscientific account, theory, or belief, etc." The epigraph carries no citation, even though the "n." after the word indicates a dictionary source. None of the standard dictionaries in my office use terms like "fictitious" or "unscientific" to define "myth," and I can't help but think that at best Rothman had to look hard to find a definition of the term suitable to his argument that "myth" is a genteel equivalent of "lie."

Of course "myth" is the substance in one form or another of literature. *The American College Dictionary* defines the term first as "a traditional or legendary story, usually concerning some superhuman being or some alleged person or event, whether without or with a determinable basis of fact or a natural explanation."[3] The essence of the argument concerns—as we have seen all along, and particularly in our Foucault discussion—whose lie/story it is. While literature people are the last categorically to deny any possibilities of truth or meaning, scientists of Rothman's persuasion define themselves as exclusionists eschewing any stories that don't fit their methodology. In a way, narrowly defining admissibility provides science with a necessary rigor and control, even while it is the hallmark of theoretical cleansing. Myths for people in my business are sources of explanation, human history, creativity, and possibility. Science works with what, narrowly defined, *is*, while literature works with what *might be*. In keeping with the dialogical character of the book, we will offer our opinions serially, by chapter and "myth."

Myth: "Nothing exists until it is observed."

Bowen

Those who have followed our discussion of Plato's allegory of the cave should recognize the inherent flaws in Rothman's misinterpretation of Platonism. Rothman tells us "The Platonists of Greece rejected the outer real world in favor of the self-created inner world of pure thought" (20). We have already seen that the ideal world exists outside the cave dwellers' perception—behind them. Their confinement against the wall forces them to see only the shadows created by the light outlining the objects to be perceived, rather than the objects themselves. The argument involves a confinement that dictates the conditions of reality perception. By extension, any rigorous set of standards such as scientific laws, paradigms, methodology, and instruments are the tools of confinement, shaping the perception of those subject to it. Whether scientists see through a glass darkly, as the biblical passage indicates, or a microscope lightly, perceptions are bounded by instruments, methodology, and strict ideology. Such exclusionary tactics as those discussed above could be considered the modern equivalent of the confinement of Plato's cave crowd. It's an efficient system, but hardly the basis for a claim on all ultimate truth.

Rothman's "laws of permission and denial" (34) certainly have a legalistic tinge to them. Framed in such legalistic Old Testament terminology, it recalls our earlier discussions of scientific legalisms. But, as David has already acknowledged, when such legalisms show signs of fallibility, the paradigms are changed and old laws give way to the modification of new laws, nowhere more apparent than in quantum theory. But of course, if you followed our earlier discussions closely you will see that the acknowledgment of the falsity of old laws and the adoption of more plausible laws and paradigms is frequently cited as the cornerstone of scientific progress. If laws of permission and denial were taken as seriously as Rothman would like to think, then there would be no major "revolutions" or quantum (no pun intended) leaps forward. Instead of admitting, as even David does, the possibility that existing scientific paradigms might someday prove fallible, Rothman simply relies heavily on semantic naming games:

> *And the sign of* [the electron's] . . . *charge is by definition always negative.* If a particle turns up that is just like an electron, except that its charge is positive, then we call it something else.

As a matter of fact, there do exist particles exactly like electrons, but with positive electric charges. But we don't call them electrons—we call them positrons. (43)

Thus, if a law doesn't fit all cases, as science would have it do, simply change the terminology.

I share with Rothman and David a belief in the possibility that human processes of thought might be understood in terms of the mechanics of particle theory, synaptic connections, and so on. But I would like to keep open the door of what that means, and I wonder about the detrimental effects that could arise from manipulation of such knowledge in a system that eschews any standards of value judgment other than verifiability. When Rothman blandly asserts, "we are now much more certain about what we know and what we don't know," that was as true of the Spanish Inquisition as it is of today's scientists. We have only begun to evolve.

Wilson

Rothman's "myth" that nothing exists until it is observed, does express an opinion that has been held by some. It is a very egocentric view of reality, which dismisses the idea of a world existing independent of the observer. There are some, such as George Berkeley, who argued that "external objects" actually exist only in and for consciousness. Existence was meaningless except as it related to consciousness.

I have been taking the opposite view, that the world is the only reality, and our conscious experiences but a part of it. If a large asteroid were to impact the earth tomorrow and obliterate all human life, I accept that the rest of the universe would continue on, unobserved by us, hardly changed by the event. But I also think that it would be a profound tragedy of unimaginable proportions, especially if we are the only thinking beings in this universe.

Zack argues that scientific laws and instruments are tools of confinement. While I can agree that scientific laws can be used to specify what is possible in the world, I cannot view scientific instruments in the same way. These, from my perspective, actually offer the opposite of confinement. As we have discussed earlier, our perceptual abilities, our senses, are limited—they are a narrow window from which to view the world. Scientific instruments actually open new aspects of the world to view, and there is nothing that limits the instruments that might be built, other than reality itself.

Furthermore, the restrictions to be found in scientific laws are not imposed by the laws—that would be reading things backwards. Instead, it is the observations and experiments that have suggested to us that such restrictions exist (take those described by the second law of thermodynamics, for example), and the theories and laws of science merely express our current knowledge of such *real* restrictions. The actual restrictions (as opposed to our laws describing them) we presume are fixed (although even that is questionable and testable), while our theories are subject to change as they confront new tests—as they confront reality, so to speak.

When something new is observed, such as positively charged particles that otherwise are like electrons, we do develop a new name, positron in this case. But we do much more. Our theories are changed as well. Interestingly, in the case of positrons, the theoretical prediction of their existence actually predated their observation. Thus, in this case, rather than science merely changing the terminology, the scientist P. A. M. Dirac was ahead of the game, and predicted the existence of antimatter in general, and the positron in particular, before the demonstration of the existence of positrons.[4]

Numerous examples of the above—theory preceding observation—lend support to the idea that we should be confident about some of our scientific knowledge, although I would disagree with Rothman about how much of it.

Myth: "Nothing is known for sure."

Bowen

The second chapter is largely a defense of the evolving theories of physics dealing with explanatory designations for causes of phenomenal occurrences that were not necessarily visible to the naked eye, or the telescope, or even the early microscope. Reality for Rothman begins in particles—those we can't see as well as those we can—which, combined, form everything we can see, like rocks, plants, animals, whatever. We apply theoretical systems, including mathematics, to our understanding of the action of the particles and we achieve the propensity for understanding the universe. When the actions of the particles are hard to get hold of, we have to theorize and come up with Kuhnean paradigms that explain what it all means. If their actions seem uncertain, as Heisenberg pointed out, we have to develop means of measurement to make the necessary observations and modify or change the theories, few of which have an active shelf-life of more than fifty years, until the majority of questions are answered with a fair probability of verification by an ever-evolving machine technology we make increasingly available for the purpose.

Wilson

If "nothing is known for sure," what about the statement itself? If it is taken as true, there is a self-contradiction, because then something is known, namely, that nothing is known.

However, beyond that self-contradiction, I tend to side with Zack on some of his points. While I can understand the frustration of scientists who see others questioning what to them is very certain knowledge, I do not believe that the correct reaction is to build our fences higher. I like the word "theory" because it indicates a lack of full certainty that is appropriate for science. Ours is always an unfinished enterprise. I have earlier indicated that I do believe we have some knowledge that we can be certain of beyond reasonable doubt; that doesn't mean beyond any doubt. Some of our theories, and even laws, will undergo revision. That's science.

Myth: "Nothing is impossible."

Bowen

This chapter constitutes a recital of all the things that people cannot do because those activities violate scientifically established principles. The chapter has a ballad-like structure, with each stanza another crackpot idea, like UFOs hovering indefinitely above the ground, squaring the circle, and so on, all punctuated with the oft-repeated refrain of the perpetual motion machine, Rothman's apoplexy-producing bête noire. With border areas in which science has established its own scarcely credible parameters, such as the absolute maximum speed limit of light and the time warp as we approach that speed, science has hypothesized its own fantasy universe that sounds strangely like metaphysical speculation.

His saving rationale at the end of the chapter is that if what we don't "know" contradicts what we do "know," then we are talking about "miracles," with which Rothman particularly equates parapsychology. He concludes,

> While parapsychology has spent the past century trying to prove miracles with no results that anyone outside the parapsychology community can agree upon, physics has spent the past century building a *colossal structure of knowledge* [italics mine]. The learning curve is clearly in favor of physics. (91)

What Rothman calls parapsychology is the descendant of mythic belief systems that have for millennia been a part of human culture. The past century is certainly a brief learning curve in terms of human history. Rothman is caught in his own time warp. A century is a millisecond interval in even a human—let alone the "big bang"—time frame of history. It is hard for humanists to believe that a mere century of speculation built a "colossal structure of knowledge." The salient aspect of this to a student of imaginative literary language is Rothman's use of a double-edged metaphor of vanity and size, as if the sum total of scientific knowledge resembles an Egyptian tomb or a medieval gothic cathedral as a monument to man's vanity. In Shelley's "Ozymandias," the closing lines describe an epitaph on the sand-obliterated ruins of what had been a colossal structure:

> "My Name is Ozymandias, king of kings;
> Look on my works ye Mighty and despair!"
> Nothing beside remains. Round the decay
> Of that *colossal* wreck, boundless and bare
> The lone and level sands stretch far away. [Italics mine]

How's that myth for a concluding homily?

Wilson

I've been a fan of "Ozymandias" ever since first reading it when I was in high school. I appreciate its cautionary note concerning hubris. However, there certainly are impossibilities in our world, and I suspect that even Zack would admit to a few.

Among the items on Rothman's list, I would not have chosen to question the same ones as Zack. Zack may characterize the "speed limit" of the velocity of light as "scarcely credible" more out of ignorance concerning physics than from solid evidence. Zack also questions how confident scientists can be about knowledge we have held for a mere century or less. Confidence in particular aspects of scientific knowledge stems from the supporting evidence, not from the number of years that have passed since a hypothesis or theory was first proposed. However, I do have more caution than Rothman about claiming certainty. At the end of the nineteenth century, it appeared that the science of mechanics was quite certain and nearly complete. There were just a few mopping-up exercises to be done, such as those related to a bothersome aspect of black-body radiation and to the photoelectric effect. Then, by early in the twentieth century, Max Planck and Albert Einstein had suggested solutions to these puzzles that opened a whole new perspective on matter and on mechanics. While much of what Rothman lists will probably prove to be correct, there could be new observations that would overturn one of his claims, however small the possibility. At the same time, to believe that anything is possible is hopelessly naive.

Myth: "Whatever we think we know now
is likely to be overturned in the future."

Bowen

Rothman's argument here lies in his claim for the permanence of most modern scientific belief in contradistinction to earlier, less rigorously scrutinized, theories and opinions. His current quarrel is with science fiction buffs who claim that modern scientific theories "are but temporary abstractions," to be supplanted in the future. Rothman's offers a priori evidence that this is not true: "The reason is, as we have shown, that we do know some things for a certainty" (94). In answer to the unspoken question, "What things and why such certainty?" Rothman makes these startling assertions: "All progress in science has been an outcome of realism. By contrast, there has been no accumulated knowledge based on idealistic theories" (98). Let's close down all those wasteful libraries! What do we have in them except idealistic theories only slightly tempered by realism? If every bit of idealistic knowledge collected before and during the recent reign of modern science is worthless, surely the study of idealism, no matter how narrowly defined historically or conceptually, represents at least some smidgen of knowledge about itself, even as a component entity of Rothman's balderdash.

Building on the nineteenth-century premise that truth lies only in what we can observe, Rothman sees a legitimate evolution to such unobservable theories as the properties of six kinds of quarks, because the new theories meet the test of being "both necessary and sufficient" (96). In other words, when physicists badly need some kind of an explanation for unmeasurable phenomena, they theorize around to find one that seems statistically to answer all their questions. This process is accompanied by an "explosive growth of instrumentation" in the kinds of paraphernalia scientists assemble to authenticate the data they need to prove their theories in the first place. It might seem to fair-minded people that making claims for observation as the fundamental criterion of scientific validity for any truths to be derived, and then, lacking any means of making those observations, substituting the newly minted criteria of being "necessary and sufficient" as justification for any theory that seems to work, is changing the rules of the game. It smacks of the very indeterminacy and lack of rigor of which Rothman accuses his adversaries across the science gap. He can't accuse all those demented parapsychologists for espousing unscrutinized beliefs on one hand, and then justify his own theories as unobserved belief systems on the other, simply by rationalizing that we need an answer to a given question and the one that apparently works must be right, at least momentarily.

Wilson

Zack seems to be especially concerned about the practice in science of claiming that entities, such as quarks, exist but have not been observed. However, in our discussion of Plato's cave, we have agreed that all observations are indirect. I do not feel uncomfortable about developing theories that appear necessary and sufficient to explain observed phenomena, which is what such theories of quarks are all about.

Perhaps Zack will feel a bit more comfortable if he considers the history of the idea of atoms. While the idea that all matter is composed of atoms is many centuries old, more concrete, modern theories of atoms grew out of the study of elements and how they formed compounds. The constant, small-number ratios of the proportion of elements in various compounds certainly supported the idea that atoms existed. That idea was fleshed out early in the twentieth century as we learned about atomic nuclei, with protons and neutrons, surrounded by clouds of electrons. Such models of atoms developed from observations, and I dare to say that, by the middle of the twentieth century, all scientists believed in the existence of atoms, some years before direct "observation" of them by special microscopes became possible. I have no problem with this kind of prediction of entities. It is a question of matching theory with observation, so there is no real change of the rules of the game.

As to Rothman's assertion, I can agree that much of what we know today will not be overturned in the future. However, exactly what will prove to be right and what will be overturned (or modified), as well as what will replace it, is unknown.

Myth: "Advanced civilizations on other planets possess great forces unavailable to us on earth."

Bowen

This brief chapter elaborates on the belief that modern humanoid scientists have uncovered laws endemic in the nature of things that apply not only to our local environment but to the entire universe. The arrogance of universal assertions has hardly proved effective in local (Western) history, in which men have always tried to impose universality on their most cherished mythologies. Attempts at scientific imperialization of outer space jurisprudence to conform to our own beliefs in universal laws is like the Jesuits invading the New World, confident in the absolute applicability of their own belief system to the rest of the world. But what apparently works in Europe doesn't necessarily have to apply to the Caribs or the Aztecs. It is part of the human psyche to project onto other persons and other environments our own way of thinking, but to assume that it can possibly approach truth for everyone or every thing in every environment in the universe is a solipsism that flies in the face of human history.

Universal physical laws are theoretical constructs that may work for us now, but as scientists themselves tell us, they are subject to change as science learns more. They come from men, not God or some natural environmental equivalent, but that they will remain intact everywhere forever approaches religious (to use a derogatory term for many humanists as well as scientists) speculation.

Rothman's immediate application of the incontrovertibility of scientific laws is the old UFO chestnut, with its energy fields at variance with current legislation. Rothman cites one of the most creative mythic theory conjunctions, that of particle theory and its application to cosmic rays—perhaps now metamorphosed by Rothman into a "principle"—as proof of the uniformity of all space. I am impressed by his ability to bounce like a particle between speculation and actuality. At any rate he and the rest of us are "left with uniformity: the cosmos is one. It is uniform in composition throughout all space" (121). Even though the statement has the ring of coming from Our Ford, we can only reasonably answer, "Maybe."

Wilson

In the end, the accuracy of any scientific claim to know about the rest of the universe, outside of our local area, must rest on observation. As Rothman describes, we do have some knowledge of what the rest of the universe consists of, from the light reaching us from distant stars and galaxies. I don't think we claim that everything is uniformly the same, but there is good evidence that our sun is not an unusual star, and more recently we have gained direct evidence of the existence of planets around other stars. Far from solipsism, we also have identified objects in the universe that are not to be found in our own solar system. Supernovas, neutron stars, and black holes are among these objects, and hypotheses concerning the details of their properties have been proposed. In the data comes evidence for the general uniformity of the universe, expanding from a singular event twelve billion years ago, but not for everything being like our own solar system.

It may seem amazing to Zack that we can know about objects so far away from us, but distance alone is not the proper measure of ignorance or knowledge.

Myth: "All scientists are objective."

Bowen

This chapter opens with a pop-culture stereotype approach to the characterization of scientists. Many of my funny, warm-and-cuddly, introspective, brilliant friends (including David) are scientists, so I really can't buy into Rothman's characterization as having any remote application to stereotypical behavior. While I can honestly detect in myself a smidgen or two of jealousy at scientists' comparative success in the public mind, academe, and the funding arena, I wonder if the stereotypical image applied to popular films is not somewhat defensively skewed. For every derogatory cinematic image of satanic scientists taking God's law into their own hands, I can think of others depicting hard-working scientists cooking up life-saving vaccines, preventing all sorts of world-threatening microbes, and thwarting malicious outer-space beings and conspiratorial plots from achieving universal destruction.

On the other hand, the image of humanists and literary people as forgetful, crotchety old humbugs, with desks buried in tomes and papers, pedantically pontificating at the drop of a hat on subjects few students or the general public care about, is the one that greets me every morning as I shave. The point is that stereotyping is not uncommon for any of us, regardless of our profession or discipline.

Rothman's defense of physical science occasionally reveals a tinge of paranoia when it comes to funding:

> Is it possible that some of the resistance to the superconducting supercollider project originates in a reluctance by non-scientists to probe too deeply in the fundamental structure of matter? (163)

Of course it is possible, even if such knowledge doesn't bring us once more to the brink of annihilation somewhere down the line. But a more reasonable explanation might be that the inordinate cost of the project might provide the poor with better health care, adequate housing, food, and better education.

Wilson

I happen to share Rothman's disappointment that the superconducting supercollider did not get built because I think that the new insights it would have given us about matter could have been very important. However, I also agree with Zack that there are many items on our social agenda, such as health and education, that are in need of more adequate funding.

Myth: "All problems can be solved by computer modeling."

Bowen

This myth is inspired by the opposition to animals as subjects for scientific experimentation. If animal activists actually did voice such myths in their perennial wars against the mutilation and slaughter of animals in laboratories, I doubt that such an allegation can, at least in the current state of the technology, be seriously made. But anyone who believes in the improvement of computer technology could certainly extend that belief eventually to include all the variables Rothman cites in his defense of animal testing. The problem is how many people will suffer and die in the interim because of our failure to use the animal-testing expedient before clinical testing on humans. One of the burdens that scientists will have to bear for their contributions to human knowledge is that criticism. I have no answer for the problem, but Rothman's polemic proves that many scientists, even if they are incensed by the hindering of their experimentation, are at least aware that a real problem exists, and that they can't entirely disregard the abuses they inflict on animals in the process. The animal activists have done all of us a real service in pointing out the horrors that have occurred in scientific laboratories insensitive to the cost of animal experimentation. They have produced photographs of the atrocities, and caused a significant body of legislation to safeguard animal rights, without sacrificing living species needlessly or abandoning research on the afflictions that affect people as well as animals. Decisions have to be made on the basis of the greater good for all life. The development of new cosmetics through animal torture hardly compares with investigating potential cures for cancer. Many biological scientists have a greater claim on the common good than hunters who heedlessly blow away birds and animals for sport, or fishermen who kill rapidly depleting species and then abandon their carcasses, all equally suspect in the eyes of those who revere life. The learning process is slow, but those of us, like Rothman, who believe in the myth called nature and regard life as its most vital segment have got to become aware of the cost of what we do.

Rothman elaborates on the computerization process and its adaptability to the multiplicity of variables in a variety of investigations, including those into the human thought processes, but even exclusionistic Rothman does not wholly reject the possibility of computer modeling becoming one of the more feasible methods of a greater number of problems in the future, although animal experimentation is more expeditious in other vital medical areas today.

Wilson

The vast majority of scientists are very sensitive to the need to care for animals and to minimize any necessary suffering, just as are the vast majority of pet owners. I think that Zack has been taken in a bit by the animal-rights propaganda. As a biologist, I have been surrounded by researchers doing studies on animals for almost forty years, thirteen of them with biomedical researchers at a medical school. During that time the care of animals has improved, just as has the care taken with human subjects of research. Today, researchers must submit their research plans to animal-care committees who must approve any experiments involving vertebrates.

Zack acknowledges that the use of animals in biomedical experiments at least speeds us to answers concerning medical problems, and thereby reduces human suffering. I do not think that most animals, apart perhaps from some other primates, have rights, but humans do have responsibilities toward animals. We have an obligation to treat them as humanely as possible. According to some animal-rights activists, anyone who is not a vegetarian, or who is a fisherman, or who has pets, or who thinks that a human life is more important than that of a rat, is wrong. I am not surprised that such extremists also would call for an end of the use of animals in research. The damage comes if other, more reasonable, individuals support the extremes of the movement.

Computers will not, in the foreseeable future, be able to replace animals in biomedical experiments. Let me give a simple example. If one wants to test the safety and efficacy of a new drug, and expects to be able to just plug the chemical structure of that drug into a computer and get an answer, that computer will have to have the complete structure of all of our proteins and other body molecules, as well as a complete model of how we are organized, from cell organelles, to cells, tissues, organs, and organ systems. There would have to be a complete program for all the ways that we can metabolize this new compound, as well as an analysis of how each of the partial breakdown products will affect each of the molecular structures to be found in us. If you are beginning to think that, by the time we have a computer that has all of this information, we will already have cured all human diseases, you probably are right.

Even the idea that cells in culture can completely replace animals is grossly oversimplified. For one thing, the cells in cell culture come from animals. Second, the idea fails to take into account the complex interactions among different kinds of cells that occur in animals, including humans. If a liver metabolizes a compound, the new product might be a poison to us. That breakdown product might not even be present in cell culture, and the cell-culture studies would indicate everything is OK.

Of course, the use of cell culture and computers can cut down on the number of animals used in research, and developing alternatives to animals for some aspects of experiments is an important effort. The National Institutes of Health currently is giving grants to encourage the development of new alternatives to animal testing, but the realistic goal is to reduce the number of animals used in research. There is no expectation that we will do without animal testing altogether. There is no substitute for the use of some animals in biomedical testing and experiments, nor will there be any time soon, much as we might wish otherwise.

Myth: "More technology will solve all problems."

Bowen

Essentially an essay on overpopulation, Rothman's chapter emphasizes exponential population growth in relation to ever-decreasing resources, and sees attempts to correct such immediate problems by technological means as productive of still more problems. The same applies to science's success in prolonging human life, for which the inexorable problems of aging loom ever larger with the deterioration of the physical and mental processes. Here is one place Rothman's "can't do" philosophy works for me. His solution lies in the limitation of population growth, a philosophy with which it is difficult logically to disagree.

Wilson

Of course, Rothman's point that more technology cannot be expected to solve all problems, should be an obvious one, especially to those who fear unintended side effects of new technologies. Rothman uses population growth as a counter example, and Zack and I seem to have little disagreement between us on this point.

But Rothman also goes out on a limb in recommending that we cut off research on aging that might result in life prolongation, at least until we have solved the problem of Alzheimer's disease and late-life dementia. He has made some oversimplified assumptions. Two of these assumptions are that general research into aging and life extension will not help us to solve Alzheimer's disease, and that, as we prolong life, Alzheimer's will necessarily increase in frequency. Just because the frequency of Alzheimer's disease increases with age today does not mean that the same percentage of people who have Alzheimer's at age 80 will continue to hold in the future, if people remain healthy in other ways and continue to use ("exercise") their brains. During the twentieth century, life expectancy among the elderly in the United States has increased. It appears that most of that increase has been in healthy years rather than in years spent in a diminished or disabled state. I would not want to stop trying to extend that because of an oversimplified view.

Myth: Myths about reductionism

Bowen

This series of myths involves Rothman in rationalizing his reductionist theory that "everything in the universe is composed of elementary particles interacting according to known laws—*and nothing else*" (232). His operative definition of "Reductionism" is "any method or theory of reducing data, process, or statements to seeming equivalents that are less complex or developed, usually a disparaging term" (238). In his use of this charged term, Rothman enters the semantic battle over arbitrary signifiers, which we discussed in our chapter on language and indeterminacy. The term "reductionism," itself an arbitrary signifier with no absolute meaning, assumes for Rothman a life of its own, both in the lexicon of philosophy and in the everyday pursuit of science in the laboratory.

The habit of thinking of particles as the real building blocks of matter is now firmly ingrained in the psyche of physicists. Using the
equations of theoretical physics to predict how these particles are
going to behave is the only way that physicists know how to think.
When critics and philosophers discuss how physicists think, they
call it reductionism. But physicists hardly ever stop to think they
are engaging in reductionism. (223)

If theirs is not to reason why, Rothman will do that for them. He traces
the term through the twentieth century, with its derivative corollaries,
from *methodological reductionism* (a description of research techniques
borrowed from one scientific discipline and applied to research problems
of another discipline) to *ontological reductionism* (appropriated from
philosophy, to deal with the metaphysical aspects of "being, reality, or
ultimate substance"). Rothman reiterates the latter to be "nothing more
than the basic premise of . . . [this] book: the universe is composed of
elementary particles and the forces by which they interact, *and nothing
else* [emphasis Rothman's]" (222).

Having proclaimed his stance as reductionist, Rothman describes "lin-
earity," which traces a lockstep chain of events "from ground up," and
"holism," which examines the totality of a complex process "from the top
down." He sees the serial passage of information-charged particles culminating in some kind of systematic translation which makes holistic
sense. The process applies to computers and everything else, from the
remote control for the TV to the human mental complexities of writing
The Divine Comedy. Obviously stung by the reviewer who equated
his purely mechanical reductionism with atheism, Rothman turns his
reviewer's commentary around to accuse him of harboring dark religious
beliefs. "I think that in many other cases, criticisms of reductionism come
from writers attempting to be scientific while a religious soul is pushing
out from behind the curtain. And that is the hidden agenda" (234).

That suspicion unequivocally announced, Rothman goes on to iterate
once again his own negative seven-fold version of the Ten Command-
ments: "Nobody is ever going to hang levitated between floor and ceiling
. . ." (236), and he closes with a final admonition: "It should be easy for a
young scientist to decide where to put his time and effort" (237). Certainly not in literary study.

Wilson

Zack is more concerned than I about the view that we are, in the end, just atoms and forces. In science, the analysis of parts often gives us insight into the whole. We can gain great confidence in the accuracy of our descriptions at one level when we can "back them up" with a detailed understanding of what is happening at the next level down. I gave one example of this with the properties of H_2O and another when describing support for completeness in the periodic table of the elements in earlier chapters.

Of course, not all progress in science, and not all understanding, comes from reductionist analysis, but a lot does. Will atoms and forces alone be enough to explain life, or will other, vital forces be found to be necessary? That is the real question, and a reductionist approach is merely one that assumes that there are no new, vital forces involved. I think Rothman's argument will be more convincing when we are able to understand more fully how conscious experience arises from matter.

None of this is intended to diminish the importance of culture and language in shaping us or making us human. What the "reductionist" view of life suggests is that other, nonmaterial entities or vital forces do not seem to be required to understand the activities of life. Obviously, the complexities of mind require that we study it, think about it, and learn about it on a number of levels. Reductionist neurobiology is one such level; insights from literature offer another.

Myth: "Myths are just harmless fun and good for the soul."

Bowen

The first paragraph or two of his condemnation of myths in general is vintage Rothman. After generously conceding that most "Prescientific humans [which includes most living humans]" need simplistic explanations for the universe they inhabit, he descends Dante-like the ethical ladder of those who blame fate for their calamities, down to the last bolgia of people who attempt to evade the consequences of the actions by saying "the devil made me do it."

In the second paragraph, he is back on Purgatorio pointing out the peripheral benefits of mythology for developing the mini-minds of children, giving rise to the *"metamyth"* that *"myths are good for you and are also great fun,"* an idea that without passing "go" leads in Rothman's argument directly to governmental abuse and Republican Party thinking (240). While his liberal political philosophy has my unequivocal blessing, his ad hominem logic only leads him further into the dark wood of error. Having exposed "the dark underbelly of mythology" he equates its pleasures with "drinking, drug using, and sex" and the "deplorable consequences" of "their indiscriminate and uncontrolled use." All these evangelical hell-fire and brimstone admonitions stem from his sloppy ideological misuse or alternative use of the word "myth" discussed at the beginning of the course and this chapter. In reality, like "reductionism," it shows, if anything, that the most virulent results of misperceptions between signifier and signified are not confined to six-year-old Stephen Dedaluses, as we have learned in our discussion of language and perception. Had Rothman spent a smidgen of the time he devoted to physics in an English literature class his book might have been founded on far safer linguistic ground.

The balance of the chapter concerns individual behavior and its genetic, cultural, and behavioral predispositions or lack of them—much of the same material and issues we have discussed previously. The final lesson as Rothman sees it is a "new principle": "the *separation of myth and state*" (247). In short, we should not use false predictions of national vulnerability to keep the industrial-military complex thriving. In declaring our own bankruptcy along with that of the Soviets, Rothman engages in a harmless bit of hyperbole, not unlike some of my own, as any reader who has got this far in the book might surmise. But I wouldn't call a little exaggeration "mythic" even if it has been a lot of fun.

Wilson

The myths of literature and art not only give us entertainment, but also insights into human nature. How can that not be seen in our earlier analysis of *Brave New World, The Crying of Lot 49,* and *The French Lieutenant's Woman?*

Must myths be in the domain of pleasure only? No. However, it also can be dangerous to believe things that are not true, whatever their source. We also should be careful not to accept all of science and technology as good. Whether the technologies that result from science are good or bad depends upon how we, society and each of us, use them. There are risks associated with technology, but without it there would be even more suffering in the world today, and more time spent on gaining the essentials and expending effort on the more mundane aspects of life— food and shelter. One need only compare the typical life in a technologically developed nation with that in a less-developed one. Technological advances have given us more time to create and to enjoy art and literature. Have you read a good book lately?

Notes

Preface

1. With all its faults as a solitary indicator of ability, perhaps the Graduate Record Examination (GRE) is as good as any single indicator of what specific aptitudes might be prerequisite for a career in science as opposed to a literary/humanistic pursuit. The English Department looks for people with high verbal skills and is less concerned with those with quantitative skills, while the opposite is true in science departments. The intersection of both sides of the disciplinary divide lies in the analytical skills, the reasoning power that the skills generate. Unlike many previous disquisitions on the subject, this book attempts to exemplify rather than to treat abstractly the way scientists and literary scholars severally *think* about specific concerns common both to science and to literature and the humanities generally.

The conflation of literature and the whole of the humanities may be a reflection on the disciplinary hubris of our literary scholar (Zack), but a literary point of view does in a sense speak for the humanities when it reconstructs and reconceptualizes science in fictive terms that deal with the problematics and humanistic speculations about what science is all about. Certainly this approach does not represent the only truths for humanists (who write about all sorts of things not necessarily pertaining to the physical world). However, literature does devote a not insubstantial part of its work to the ideas and practices of science, even as science is forever nibbling at the parameters of issues and ideas that concern humanists.

Broadly defined, literature is not limited to poetry and fiction, but includes everything written down, from lab reports through accounts in professional and popular scientific journals, as well as philosophical and historical treatises, newspaper articles, and so on. Historically, at one time or another literature has appropriated all sorts of documents for its own disciplinary mill. This book is only another example of that acquisitive tendency.

2. One difficulty in presenting creative works to readers of the present text is how to deal with an analysis of fictional works if our readers have not read them. The difficulty is endemic in normal literature classes where assignments are discussed before or in lieu of their having been read by some of the students. What profit may

be gleaned from the discussion would be greatly enhanced by a prefatory reading of the discussed texts. Even though every attempt was made in the present discussions of passages from the fiction to explain their contexts in the assigned works, nothing provides the full context except reading the entire work itself. To that end, brief citations of the principal literary works discussed in each chapter are given in the notes to that chapter and full citations in the first section of the bibliography, "Works Cited in Text." The second section of the bibliography lists other works recommended as supplementary reading.

Chapter 1. Origins

1. For more details about the big bang and related matters, see Hawking's *A Brief History of Time* or Weinberg's *The First Three Minutes*.

2. Almost any introductory biology textbook will provide a summary of our knowledge of the origin and evolution of life. *Biology* by Campbell, Reece, and Mitchell is a good example (see the bibliography).

3. If the alternatives for the fate of the universe interest you, try *The Last Three Minutes*, by Davies, or "The Fate of Life in the Universe," by Krauss and Starkman.

4. Actually, "time" may be a problem here. The modern view is that time did not come into existence until the big bang, and so the question of what came before it is considered meaningless by some scientists.

5. Modern mythologists, heavily influenced by Jung and Neumann, have noted two coequal principles nearly universal in archetypal cyclical configurations: a masculine creative agency (Shiva) and a feminine creative agency (Inanna). The chief difference between the cosmic cycle and the nature cycle is that the first is associated with the masculine and the second with the feminine. The interaction of the two sources of creativity is one of the germinal points of psychological development of individual nature (Henderson and Oakes, 6).

For the ancients, the snake represents both a fear of death (presumably because of its venom) and the life cycle as a whole (it periodically sheds it skin). The uroboros is a metaphor for a knowledge of life forbidden in the biblical sense to any but immortals, who are outside the cycle (Henderson and Oakes, 36).

6. Henderson and Oakes, 11.

Chapter 2. How to Do Science

1. My description of what constitutes the method of science may seem hopelessly naive to some, but its practical straightforwardness serves the purpose. For more detailed introductions to scientific thinking you can try books on the philosophy of science, such as those by O'Hear or Richards.

2. If you want to know more about the explanations we have for most of the special properties of living organisms, try my *Introduction to Biology*.

3. Sir Karl Popper emphasized this in his book, *The Logic of Scientific Discovery*.

4. What follows is an example adapted from Wason's 1966 article on reasoning.

5. For a good review, see Beadle's article on biochemical genetics.

6. The original, 1905, paper by Albert Einstein is in German, but a good translation can be found in the book *The Principle of Relativity*, by Lorentz et al. In 1905 Einstein actually published five papers. The one I am referring to was entitled (in translation) "On the Electrodynamics of Moving Bodies." It has come to be known as Einstein's theory of special relativity, perhaps his best-known work. It is approachable with a little knowledge of mathematics. (More than ten years later Einstein completed his work on relativity with his publication of a theory of general relativity.) Among the other papers published in 1905 was a description of the photoelectric effect, in which Einstein concluded that light could have "particle-like" properties, a finding that became part of the foundation of quantum mechanics. It was this latter work that actually was mentioned, and not special relativity, when Einstein was awarded the Nobel Prize. Another 1905 paper was like a postscript to the special relativity paper, adding "by the way, $E = mc^2$." With this short article, Einstein identified matter as another form of energy, laying the theoretical foundation for the conversion of matter to energy in nuclear weapons and reactors. It was quite a year for a patent clerk.

7. John Platt first described this approach and called it strong inference, the title of his paper in *Science*.

8. What follows is adapted from Cromer's *Connected Knowledge*.

9. For a fuller treatment of the interaction between science and literature during the Renaissance see Ann Blair and Anthony Grafton, "Reassessing Humanism and Science."

10. Thomas Kuhn, *The Structure of Scientific Revolutions*, 167–68.

11. Jean-François Lyotard, *The Postmodern Condition: A Report on Knowledge*, 27.

Chapter 3. Can There Be a Science of Humans?

1. See his *Sociobiology: The New Synthesis*.

2. Beyond the nutshell, try the original: Darwin's *The Origin of Species by Means of Natural Selection*, or any one of a number of more modern books that update the topic with new insights. A popular recent source would be E. Mayer's *This Is Biology*.

3. If you need convincing, try *The Blind Watchmaker*, by Dawkins.

4. See Jensen's paper in the *Harvard Educational Review*.

5. See Flynn's 1984 and 1987 papers in the *Psychological Bulletin*.

6. For those who would like a taste of other counterarguments, some more strongly stated than mine, try *Not in Our Genes*, by Lewontin et al.

7. L. J. Jordanova, ed., *Languages of Nature: Critical Essays on Science and Literature*, 30–31.

8. For an explanation of the literary qualities of Darwin's writing see Gillian Beer, "'The Face of Nature,'" 207–8.

Chapter 4. Darwinism, and Knowledge and Power

1. Foucault, *Power/Knowledge*, 78–108.
2. Weston, "The Virtual Anthropologist," 168.

Chapter 5. Darwin and *The French Lieutenant's Woman*

1. Fowles, *The French Lieutenant's Woman*, 23.
2. Kaplan, 111.

Chapter 6. Uncertainty

NOTE: A version of the first part of this chapter has previously been published in *Explorations* 17 (1999) and is used here by permission of the author.

1. For those interested in what has been said, and in some strong reactions to it, try *Higher Superstition* by Gross and Levitt.
2. Laplace, in 1812, put forward a very clear statement of determinism with his concept of a Divine Calculator, who, knowing the position and velocity of all of the particles in the universe, could determine the future and past of the entire universe. Mason's *A History of the Sciences* gives a succinct review on page 296.
3. John Earman discusses this issue and many others in his *Primer on Determinism*.
4. For more detail there are many sources available on the subject. James Gleick's book is a popular exposition on the topic.
5. My introduction to the subject came from Eisberg's *Fundamentals of Modern Physics*, which presumes a knowledge of calculus and some physics. There are any number of books available on quantum mechanics at various levels of sophistication.
6. A paper of mine on this subject appeared in the *Journal of Consciousness Studies*, 1999.
7. Gazzaniga, in *The Mind's Past*, takes a strong position of this sort.
8. See my 1976 article in *Perspectives in Biology and Medicine*.
9. Damasio's *Descartes' Error* offers a fascinating presentation of the evidence.
10. *The Portable Nietzsche*, 46–47.
11. *Manifesto of a Passionate Moderate* (1998).
12. I am indebted to Leonard Orr's synopsis of these schools of thought in his *Dictionary of Critical Theory*, 358–65, 369–99, 105–15.
13. *Collected Papers*, 1: 339.
14. Saussure, *Course in General Linguistics*, as quoted in Orr's *Dictionary of Critical Theory*, 462.
15. As, for example, Frazer, in *The Golden Bough*, and Durkheim and Mauss, in *Primitive Classification*.
16. M. Gell-Mann, in a private letter to the editor of *The Oxford English Dictionary*, June 27, 1978. It appears in the *OED*, 2d ed., 12:984.

Chapter 7. Preparing for Pynchon

1. For those who know some science and are interested, the book, *Maxwell's Demon: Entropy, Information, Computing,* by Leff and Rex, nicely covers the history of the ideas.

2. If you want further insight into the difference between information and meaning, try John Searle's Chinese room argument, in chapter 2 of his *Minds, Brains, and Science.*

Chapter 8. Thomas Pynchon's *The Crying of Lot 49*

1. Lance Olsen, "Pynchon's New Nature: The Uncertainty Principle and Indeterminacy in *The Crying of Lot 49," Canadian Review of American Studies,* 4, no. 2 (1983), 156.

2. I am grateful to Michael Gillespie for the conjunction of Monroe and Echo.

Chapter 9. Perception and Reality

1. Trans. Francis MacDonald Cornford, 227–35. See the bibliography for full reference.

2. See Homi K. Bhabha's "Signs Taken for Wonders," 24–28.

Chapter 10. Revolutions in Science

1. Pope John Paul II made his statement on evolution in an address to the Pontifical Academy of Sciences on October 22, 1996. It has been reprinted in *The Sciences,* and a copy is on the web (see John Paul II in the bibliography).

2. See, for instance, John Haught's *Science and Religion,* which speaks of traditional religion picturing the universe as a hierarchy of these "distinct" levels (72–73).

3. For instance, Edelman describes some of the impacts that a neuroscientific revolution could have in his book, *Bright Air, Brilliant Fire.*

4. See, for example, T. H. Huxley's *Method and Results,* 244.

5. The argument that follows is rather obvious, once one has seen it. I thought I might have been the first with the idea in my paper of 1976, but later found that H. J. Muller had made a similar argument in 1955, and gave him credit in a paper I wrote in 1978. I would not be surprised if the idea occurred to someone even earlier than 1955.

6. A modern statement of this view can be found in the Carins-Smith book, *Evolving the Mind.*

7. Three recent books give different perspectives on how data from neuroscience is supporting monistic views: Crick's *The Astonishing Hypothesis* examines evidence from studies of how the brain processes visual input; Hobson's *Consciousness* gives insights from studies of sleep and dreaming; and McCrone's *Going Inside* describes data from brain imaging studies, which can tell us what brain regions are active when we are thinking particular thoughts.

8. See Paul Churchland's *The Engine of Reason, The Seat of the Soul,* 7–8.

9. See Gazzaniga, Bogen, and Sperry, 1962, for one of the original papers. A good summary can be found in Sperry's 1968 review. One of many books written on the subject is *The Integrated Mind,* by Gazzaniga and LeDoux.

10. The quotations are from the copy of the pope's address found at the New Advent Catholic website (http://www.knight.org/advent/docs/jp02tc.htm).

11. Joseph Heller, *Catch 22,* 441–42.

Chapter 11. Cycles and History/Yeats

1. In a sense what David and I adopted as the format of this book was what Bakhtin would call "dialogical," a dialogue derived from two opposing points of view or narrative perspectives. Hopefully the desired result would be some sort of synthesis in the reader's understanding of both perspectives, how they differ and what they have in common in the pursuit of knowledge.

Chapter 12. Ethics and Morality in Science and Technology

1. For a detailed analysis of the acceptance here of former Nazi scientists, see Michael J. Neufeld, *The Rocket and the Reich.*

Chapter 14. Summary

1. Tennessee Williams, *A Streetcar Named Desire.*

Appendix B. The Science Gap

1. See William J. Broad, "Paul Feyerabend: Science and the Anarchist," in *Science.*

2. London: New Left Books, 1975 and 1978, respectively.

3. New York: Random House (1968), 805.

4. Read about it in Dirac's own words in his book, *The Development of Quantum Theory.*

Bibliography

Works Cited in Text

Beadle, George. "Biochemical Genetics." *Chemical Review* 37 (1945): 15–96.

Beer, Gillian. "'The Face of Nature': Anthropomorphic Elements in the Language of *The Origin of Species*." In *Languages of Nature: Critical Essays on Science and Literature*, ed. L. J. Jordanova, 207–43. New Brunswick, N.J.: Rutgers University Press, 1986.

Bhabha, Homi K. "Signs Taken for Wonders," In *The Post-Colonial Studies Reader*, ed. Bill Ashcroft, Gareth Griffiths, and Helen Tiffin. London: Routledge, 1995.

Blair, Ann, and Anthony Grafton. "Reassessing Humanism and Science." *Journal of the History of Ideas* 53 (1992): 535–61.

Broad, William J. "Paul Feyerabend: Science and the Anarchist." *Science* 206 (November 2 ,1979): 534–37.

Campbell, Neil, Jane Reece, and Lawrence Mitchell. *Biology.* 5th ed. Menlo Park, Cal.: Benjamin Cummings, 1999.

Carins-Smith, A. G. *Evolving the Mind: On the Nature of Matter and the Origins of Consciousness.* Cambridge: Cambridge University Press, 1996.

Churchland, Paul M. *The Engine of Reason, the Seat of the Soul: A Philosophical Journey into the Brain.* Cambridge, Mass.: MIT Press, 1995.

Crick, Francis. *The Astonishing Hypothesis: The Scientific Search for the Soul.* New York: Charles Scribner's Sons, 1994.

Cromer, Allan. *Connected Knowledge: Science, Philosophy, and Education.* New York: Oxford University Press, 1997.

Damasio, Antonio. *Descartes' Error: Emotion, Reason, and the Human Brain.* New York: G. P. Putnam's Sons, 1994.

Darwin, Charles. *The Origin of Species by Means of Natural Selection.* New York: Humboldt Publishing Co., 1890.

Davies, Paul. *The Last Three Minutes.* New York: Basic Books, 1994.

Dawkins, Richard. *The Blind Watchmaker.* Longman, 1986.

Dirac, P. A. M. *The Development of Quantum Theory.* New York: Gordon and Breach, 1971.

Durkheim, Emile, and Marcel Hauss. *Primitive Classification,* trans. Rodney Needham. Chicago: University of Chicago Press, 1963.

Earman, John. *A Primer on Determinism.* Boston: D. Reidel, 1986.

Edelman, Gerald. *Bright Air, Brilliant Fire: On the Matter of Mind.* New York: Basic Books, 1992.

Eisberg, Robert. *Fundamentals of Modern Physics.* New York: John Wiley and Sons, 1961.

Feyerabend, Paul. *Against Method.* London: New Left Books, 1975.

———. *Science in a Free Society.* London: New Left Books, 1978.

Flanagan, Owen. *Consciousness Reconsidered.* Cambridge, Mass.: MIT Press, 1992.

Flynn, John. "The Mean IQ of Americans: Massive Gains." *Psychological Bulletin* 95 (1984): 29–51.

———. "Massive IQ Gains in 14 Nations: What IQ Tests Really Measure." *Psychological Bulletin* 101 (1987): 171–91.

Foucault, Michel. "Two Lectures," *Power/Knowledge: Selected Interviews and Other Writings, 1972–1977.* Edited by Colin Gordon. New York: Random House, 1980. 78–108.

Fowles, John. *The Collector.* Boston: Little, Brown: 1997.

———. *The French Lieutenant's Woman.* New York: Penguin Books, 1970.

Frazer, James G. *The Golden Bough.* 12 vols. London: Macmillan, 1907–15.

Gazzaniga, Michael S. *The Mind's Past.* Berkeley: University of California Press, 1998.

Gazzaniga, Michael S., J. E. Bogen, and R. W. Sperry. "Some Functional Effects of Sectioning the Cerebral Commissures in Man." *Proceedings of the National Academy of Sciences, U.S.* 48 (1962): 1765–69.

Gazzaniga, Michael S., and Joseph E. LeDoux. *The Integrated Mind.* New York: Plenum Press, 1978.

Gleick, James. *Chaos: Making a New Science.* New York: Viking, 1987.

Gould, S. J. *The Mismeasure of Man.* New York: W. W. Norton, 1981.

Gregory, R. L. *Eye and Brain.* New York: McGraw-Hill, 1966.

Gross, Paul, and Norman Levitt. *Higher Superstition: The Academic Left and Its Quarrels with Science.* Baltimore: Johns Hopkins University Press, 1998.

Gupta, Akhi, and James Ferguson. *Anthropological Locations: Boundaries and Grounds of a Field Science.* Berkeley: University of California Press, 1997.

Haack, Susan. *Manifesto of a Passionate Moderate.* Chicago: University of Chicago Press, 1998.

Haught, John F. *Science and Religion: From Conflict to Conversation.* New York: Paulist Press, 1995.

Hawking, Stephen. *A Brief History of Time.* New York: Bantam, 1988.

Heller, Joseph. *Catch 22.* New York: Dell Publishing, 1961.

Henderson, Joseph L., and Maud Oakes. *The Wisdom of the Serpent: The Myths of Death, Rebirth, and Resurrection.* New York: George Braziller, 1963.

Herrnstein, R. J., and C. Murray, *The Bell Curve: Intelligence and Class Structure in American Life.* New York: Free Press, 1994.

Hobson, J. A. *Consciousness.* New York: Scientific American Library, 1999.

Huxley, T. H. *Methods and Results.* New York: D. Appleton and Co., 1897.

Jensen, A. R. "How Much Can We Boost IQ and Scholastic Achievement?" *Harvard Educational Review* 39 (1969): 1–123.

John Paul II. "Truth Cannot Contradict Truth." Address to the Pontifical Academy of Sciences, Oct. 22, 1996. *The Scientist,* May 12, 1997, 8–9. Also at http/www.knight.org/advent/docs.jp02tc.htm

Jordanova, L. J. "Introduction," *Languages of Nature: Critical Essays on Science and Literature.* New Brunswick, N.J.: Rutgers University Press, 1986.

Kaplan, Fred. "Victorian Modernists: Fowles and Nabokov." *Journal of Narrative Technique* 3, no. 2 (May 1973): 108–20.

Krauss, Lawrence, and Glenn Starkman. "The Fate of Life in the Universe." *Scientific American,* November 1999, 59–65.

Kuhn, Thomas S. *The Structure of Scientific Revolutions.* Chicago: University of Chicago Press, 1962.

Leff, Harvey, and Andrew Rex, eds. *Maxwell's Demon: Entropy, Information, Computing.* Princeton: Princeton University Press, 1990.

Lewontin, R. C., Steven Rose, and Leon Kamin. *Not in Our Genes.* New York: Pantheon Books, 1984.

Lorentz, H. A., A. Einstein, H. Minkowski, and H. Weyl. *The Principle of Relativity.* New York: Dover Publications, 1923.

Lyotard, Jean-François. *The Postmodern Condition: A Report on Knowledge,* trans. Geoff Bennington and Brian Massumi. Minneapolis: University of Minnesota Press, 1991.

McCrone, John. *Going Inside: A Tour Round a Single Moment of Consciousness.* London: Faber and Faber, 1999.

Mason, Stephen F. *A History of the Sciences.* New York: Collier Books, 1962.

Mayer, E. *This Is Biology: The Science of the Living World.* Cambridge, Mass.: Harvard University Press, 1997.

Meselson, M., and F. W. Stahl. "The Replication of DNA in *Escherichia coli.*" *Proceedings of the National Academy of Sciences, U.S.* 44 (1958): 671–82.

Muller, H. J. "Life." *Science* 121 (1955): 1–9.

Neufeld, Michael J. *The Rocket and the Reich: Peenemunde and the Coming of the Ballistic Missile Era.* New York: Free Press, 1995.

Nietzsche, Friedrich. *The Portable Nietzsche,* trans. Walter Kaufmann. New York: Viking, 1954.

O'Hear, Anthony. *An Introduction to the Philosophy of Science.* Oxford: Oxford University Press, 1989.

Olsen, Lance. "Pynchon's New Nature: The Uncertainty Principle and Indeterminacy in *The Crying of Lot 49.*" *Canadian Review of American Studies* 4, no. 2 (1983): 153–63.

Orr, Leonard. *A Dictionary of Critical Theory.* Westport, Conn.: Greenwood Press, 1991.

Orwell, George. *1984.* New York: Harcourt, Brace, 1949.

Peirce, Charles. *Collected Papers of Charles Saunders Peirce,* ed. Charles Hartshorne and Paul Weiss. Vol. 1. Cambridge, Mass.: Harvard University Press, 1958.

Plato. *The Republic of Plato,* trans. Francis MacDonald Cornford. New York and London: Oxford University Press, 1963. (For the "Allegory of the Cave" in the *Republic,* see pp. 227–35.)

Platt, John R. "Strong Inference," *Science* 146 (1964): 347–53.

Popper, Karl. *The Logic of Scientific Discovery.* New York: Harper Torchbooks, 1968.

Rhodes, Richard. *The Making of the Atomic Bomb.* New York: Simon and Schuster, 1986.

Richards, Stewart. *Philosophy and Sociology of Science.* 2d ed. Oxford: Basil Blackwell, 1987.

Rothman, Milton A. *The Science Gap: Dispelling the Myths and Understanding the Reality of Science.* Buffalo: Prometheus Books, 1992.

Saussure, Ferdinand de. *Course in General Linguistics,* ed. Wade Baskin. New York: McGraw-Hill, 1966.

Searle, John. *Minds, Brains, and Science.* Cambridge, Mass.: Harvard University Press, 1984.

Shockley, William. "Dysgenics, Geneticity, Raceology: A Challenge to the Intellectual Responsibility of Educators." *Phi Delta Kappan,* January, 1972, 307.

Snow, C. P. *The Two Cultures and the Scientific Revolution.* New York: Cambridge University Press, 1959.

Sperry, R. W. "Hemisphere Deconnection and Unity in Conscious Awareness." *American Psychologist* 23 (1968): 723–33.

Terman, Lewis M. *The Measurement of Intelligence.* Boston: Houghton Mifflin, 1916.

Trefil, James, and Robert M. Hazen. *The Sciences: An Integrated Approach.* New York: John Wiley and Sons, 1995.

Wason, P. C. "Reasoning." In *New Horizons in Psychology,* ed. Brian M. Foss, 145–47. Baltimore: Penguin, 1966.

Weinberg, Steven. *The First Three Minutes.* New York: Basic Books, 1988.

Weston, Kath. "The Virtual Anthropologist." In *Anthropological Locations,* ed. Akhil Gupta and James Ferguson, 163–84. Berkeley: University of California Press. 1997.

Wilson, David L. "On the Nature of Consciousness and of Physical Reality." *Perspectives in Biology and Medicine* 19 (1976): 568–81.

———. "Brain Mechanisms, Consciousness, and Introspection." In *Expanding Dimensions of Consciousness,* ed. A. A. Sugarman and R. E. Tarter, 3–23. New York: Springer, 1978.

————. "Mind-brain Interaction and Violation of Physical Law." *Journal of Consciousness Studies* 6 (1999): 185–200.

————. "Reflections on E. O. Wilson's Views of Free Will." *Explorations* 17 (Summer 1999): 39–49.

————. *Introduction to Biology.* Malden, Mass.: Blackwell Science, 2000.

Wilson, Edward O. *Sociobiology: The New Synthesis.* Cambridge, Mass.: Harvard University Press, 1978.

————. *Consilience: The Unity of Knowledge.* New York: Alfred A. Knopf, 1998.

Supplemental Reading

Albright, Daniel. *Quantum Poetics.* Cambridge and New York: Cambridge University Press, 1997.

Berkeley, George. *Principles of Human Knowledge.* La Salle, Ill.: Open Court Publishing Co., 1940.

Bowen, Zack. *A Reader's Guide to John Barth.* Westport, Conn.: Greenwood Press, 1986.

Cristie, John, and Sally Shuttleworth, eds. *Nature Transfigured: Science and Literature, 1700–1900.* Manchester: Manchester University Press, 1989.

Dale, Peter Allan. *In Pursuit of a Scientific Culture.* Madison: University of Wisconsin Press, 1989.

Deery, June. *Aldous Huxley and the Mysticism of Science.* New York: St. Martin's Press, 1996.

Deleuze, Gilles, and Felix Guattari. *Anti-Oedipus: Capitalism and Schizophrenia,* trans. Robert Hurley, Mark Seem, and Helen R. Lane. New York: Viking, 1977.

Gabel, John B., and Charles B. Wheeler. *The Bible as Literature.* New York: Oxford University Press, 1986.

Garvin, Harry R., and James M. Heath, eds. *Science and Literature.* Lewisburg, Pa.: Bucknell University Press/Associated University Presses, 1983.

Goldberg, B. Z. *The Sacred Fire: The Story of Sex in Religion.* New York: Grove Press, 1930.

Harding, Sandra. *The Science Question in Feminism.* Ithaca, N.Y.: Cornell University Press, 1986.

————. *Whose Science? Whose Knowledge?* Ithaca, N.Y.: Cornell University Press, 1991.

Harris, Kevin. *Sex, Ideology, and Religion: The Representation of Women in the Bible.* Totowa, N.J.: Barnes and Noble Books, 1984.

Hawkins, Harriet. *Strange Attractors: Literature, Culture, and Chaos Theory.* New York: Prentice Hall, 1995.

Hayles, N. Katherine. *The Cosmic Web: Scientific Field Models and Literary Strategies in the Twentieth Century.* Ithaca, N.Y.: Cornell University Press, 1984.

Heisenberg, Werner. *Physics and Beyond: Encounters and Conversations,* trans. Arnold J. Pomerans. New York: Harper and Row, 1971.

Hooke, S. H. *Middle Eastern Mythology.* Harmondsworth and Baltimore: Penguin Books, 1963.

Huxley, Aldous. *Literature and Science.* New York: Harper and Row, 1965.

Isaacs, Leonard. *Darwin to Double Helix: The Biological Theme in Science Fiction.* London: Butterworths, 1977.

Jameson, Fredric. *The Political Unconscious: Narrative as a Socially Symbolic Act.* Ithaca, N.Y.: Cornell University Press, 1981.

Peterfreund, Stuart, ed. *Literature and Science: Theory and Practice.* Boston: Northeastern University Press, 1990.

Phillips, John A. *Eve: The History of an Idea.* San Francisco: Harper and Row, 1984.

Phipps, William E. *Genesis and Gender: Biblical Myths of Sexuality and Their Cultural Impact.* New York: Praeger, 1989.

Sarup, Madan. *An Introductory Guide to Post-Structuralism and Postmodernism.* Athens: University of Georgia Press, 1993.

Silverman, Kaja. *The Subject of Semiotics.* New York: Oxford University Press, 1983.

Strehle, Susan. *Fiction in the Quantum Universe.* Chapel Hill: University of North Carolina Press, 1992.

Van der Ziel, Aldert. *Genesis and Scientific Inquiry.* Minneapolis: T. S. Denison and Co., 1965.

Wallace, Howard N. *The Eden Narrative.* Atlanta: Scholars Press, 1985.

Index

Zack Bowen is editor of the Florida James Joyce Series and the James Joyce Literary Supplement. He is author of numerous books, including *Ulysses as a Comic Novel* (Syracuse, 1989) and *Bloom's Old Sweet Song* (UPF, 1995). He is professor of English at the University of Miami.

David Wilson is professor of biology at the University of Miami. He has published more than 50 articles in science journals.